Finite Sections of Band-Dominated Operators

Memoirs
of the
American Mathematical Society

Number 895

Finite Sections
of Band-Dominated
Operators

Steffen Roch

January 2008 • Volume 191 • Number 895 (end of volume) • ISSN 0065-9266

American Mathematical Society
Providence, Rhode Island

2000 *Mathematics Subject Classification.* Primary 65J10; Secondary 46N40, 47B35, 47B36.

Library of Congress Cataloging-in-Publication Data

Roch, Steffen, 1958–
  Finite sections of band-dominated operators / Steffen Roch.
    p. cm. — (Memoirs of the American Mathematical Society, ISSN 0065-9266 ; no. 895)
  "Volume 191, number 895."
  Includes bibliographical references.
  ISBN 978-0-8218-4042-9 (alk. paper)
  1. Operator algebras. 2. Operator theory. I. Title.

QA326.R63   2007
512'.55—dc22   2007060559

Memoirs of the American Mathematical Society

This journal is devoted entirely to research in pure and applied mathematics.

Subscription information. The 2008 subscription begins with volume 191 and consists of six mailings, each containing one or more numbers. Subscription prices for 2008 are US$675 list, US$540 institutional member. A late charge of 10% of the subscription price will be imposed on orders received from nonmembers after January 1 of the subscription year. Subscribers outside the United States and India must pay a postage surcharge of US$38; subscribers in India must pay a postage surcharge of US$43. Expedited delivery to destinations in North America US$53; elsewhere US$130. Each number may be ordered separately; *please specify number* when ordering an individual number. For prices and titles of recently released numbers, see the New Publications sections of the *Notices of the American Mathematical Society*.

Back number information. For back issues see the *AMS Catalog of Publications*.

Subscriptions and orders should be addressed to the American Mathematical Society, P. O. Box 845904, Boston, MA 02284-5904, USA. *All orders must be accompanied by payment.* Other correspondence should be addressed to 201 Charles Street, Providence, RI 02904-2294, USA.

Copying and reprinting. Individual readers of this publication, and nonprofit libraries acting for them, are permitted to make fair use of the material, such as to copy a chapter for use in teaching or research. Permission is granted to quote brief passages from this publication in reviews, provided the customary acknowledgment of the source is given.

Republication, systematic copying, or multiple reproduction of any material in this publication is permitted only under license from the American Mathematical Society. Requests for such permission should be addressed to the Acquisitions Department, American Mathematical Society, 201 Charles Street, Providence, Rhode Island 02904-2294, USA. Requests can also be made by e-mail to reprint-permission@ams.org.

Memoirs of the American Mathematical Society is published bimonthly (each volume consisting usually of more than one number) by the American Mathematical Society at 201 Charles Street, Providence, RI 02904-2294, USA. Periodicals postage paid at Providence, RI. Postmaster: Send address changes to Memoirs, American Mathematical Society, 201 Charles Street, Providence, RI 02904-2294, USA.

© 2008 by the American Mathematical Society. All rights reserved.
Copyright of individual articles may revert to the public domain 28 years
after publication. Contact the AMS for copyright status of individual articles.
This publication is indexed in *Science Citation Index*®, *SciSearch*®, *Research Alert*®,
CompuMath Citation Index®, *Current Contents*®/*Physical, Chemical & Earth Sciences*.
Printed in the United States of America.

∞ The paper used in this book is acid-free and falls within the guidelines
established to ensure permanence and durability.
Visit the AMS home page at http://www.ams.org/

10 9 8 7 6 5 4 3 2 1   13 12 11 10 09 08

Contents

Chapter 1. Introduction: Band-dominated operators	1
Chapter 2. Stability	5
2.1. Fredholm band-dominated operators and their indices	6
2.2. Algebras of matrix sequences	7
2.3. Stability vs. Fredholmness	9
2.4. Stability of the finite sections method	9
2.5. Band-dominated operators with slowly oscillating coefficients	11
2.6. The role of the index	13
2.7. The C^*-algebra of the finite sections method	14
2.8. Another perspective on localization	19
Chapter 3. Stable regularizability	23
3.1. Moore-Penrose invertibility in C^*-algebras	23
3.2. Stable regularizability vs. Moore-Penrose invertibility	25
3.3. Stable regularizability of the finite sections method	26
Chapter 4. Compactness	30
4.1. Compact sequences	30
4.2. Characterization via singular values	32
4.3. Central rank characterizations	35
4.4. Minimal and maximal characterizations	38
4.5. Compact sequences in $\mathcal{S}(\mathbb{N})$ and $\mathcal{S}(\mathbb{Z})$	41
Chapter 5. Fredholmness	47
5.1. Fredholm sequences	47
5.2. Fredholmness and stability	50
5.3. Fredholmness and stable regularizability	51
5.4. Fredholmness of the finite sections method	52
5.5. Slowly oscillating coefficients	54
Chapter 6. Essential fractality	57
6.1. Fractality of quotient maps	57
6.2. $\mathcal{J}$-fractal algebras	59
6.3. Essential fractality and singular values	61
6.4. Essential fractality of the finite sections algebras	62
Chapter 7. Applications	64
7.1. Approximation numbers	64
7.2. Rank-preserving discretizations	65
7.3. Arveson dichotomy: band-dominated operators	67

7.4.	Arveson dichotomy: the general setting	70
7.5.	Essential spectra	71
7.6.	Essential pseudospectra	73
7.7.	Pseudomodes	76
7.8.	Determinants	80

Bibliography   83

Index   86

Abstract

The goal of this text is to review recent advances and to present new results in the numerical analysis of the finite sections method for general band and band-dominated operators. The main topics are the stability of the finite sections method and the asymptotic behavior of singular values. The latter topic is closely related with compactness and Fredholm properties of approximation sequences, and the paper can also serve as an introduction into this remarkable field of numerical analysis. Further we discuss the behavior of approximation numbers, determinants, essential spectra and essential pseudospectra as well as the localization of pseudomodes of finite sections of band-dominated operators.

Received by the editor July 27, 2004, and in revised form July 14, 2005.
2000 *Mathematics Subject Classification*. Primary 65J10; Secondary 46N40, 47B35, 47B36, 47L80.
Key words and phrases. band-dominated operators, finite sections method, compact and Fredholm sequences, splitting property of singular values, Arveson dichotomy, pseudospectra.

CHAPTER 1

Introduction: Band-dominated operators

We will work on the Hilbert space $l^2(\mathbb{Z})$ consisting of all sequences $x : \mathbb{Z} \to \mathbb{C}$ with
$$\|x\|^2 := \sum_{m \in \mathbb{Z}} |x(m)|^2 < \infty.$$
The sequences e_i with $i \in \mathbb{Z}$ defined by
$$e_i(m) := \begin{cases} 1 & \text{if} \quad m = i \\ 0 & \text{if} \quad m \neq i \end{cases}$$
form a Hilbert basis of $l^2(\mathbb{Z})$ to which we refer as the standard basis. By $L(l^2(\mathbb{Z}))$ we denote the C^*-algebra of all bounded linear operators on $l^2(\mathbb{Z})$, and by $K(l^2(\mathbb{Z}))$ the set of all compact operators on $l^2(\mathbb{Z})$. Thus, $K(l^2(\mathbb{Z}))$ is a closed ideal of $L(l^2(\mathbb{Z}))$.

An operator $A \in L(l^2(\mathbb{Z}))$ with matrix representation (a_{ij}) with respect to the standard basis of $l^2(\mathbb{Z})$ is called a *band operator* if there is an integer k such that $a_{ij} = 0$ whenever $|i - j| > k$. The closure in $L(l^2(\mathbb{Z}))$ of the set of all band operators is a C^*-subalgebra of $L(l^2(\mathbb{Z}))$ which we denote by $\mathcal{A}(\mathbb{Z})$. The elements of $\mathcal{A}(\mathbb{Z})$ are called *band-dominated operators*. It is easy to see that all compact operators are band-dominated.

Needless to say that band-dominated operators occur at many places in mathematics and its applications. Typical sources of such operators are discretizations and discrete versions of differential operators. For example, the operator with matrix representation

(1.1)
$$\begin{pmatrix} \ddots & \ddots & & & & \\ \ddots & a_{-2} & 1 & & & \\ & 1 & a_{-1} & 1 & & \\ & & 1 & a_0 & 1 & \\ & & & 1 & a_1 & \ddots \\ & & & & \ddots & \ddots \end{pmatrix}$$

with a_0 standing at the 00-position is considered as a *discrete Schrödinger operator* with potential $a = (a_k) \in l^\infty(\mathbb{Z})$, and the operator (1.1) with $a_k = \lambda \cos 2\pi(\alpha k + \beta)$ with real parameters α, β and λ is known as the *Almost Mathieu operator*. A nice treatment of the spectral theory of Almost Mathieu operators and of related operators which belong to rotation algebras is [8]. See also the recent papers [5, 6] where a long-standing conjecture about the spectra of Almost Mathieu operators (the famous Ten Martini Problem) is settled. The main result of [5] says that, for irrational α and $\lambda \neq 0$, the spectrum of the Almost Mathieu operator is a Cantor set.

To mention another spectacular application: In coarse geometry, one associates with every coarse geometry space X a certain C^*-algebra $\mathcal{A}(X)$ which reflects the coarse properties of X. In case $X = \mathbb{Z}$, $\mathcal{A}(\mathbb{Z})$ is just the algebra of the band-dominated operators introduced above. For detailed accounts of coarse geometry one should consult the monographs [**24, 49, 50**].

All band-dominated operators are constituted by two kinds of basic band operators: the shift operators
$$U_k : l^2(\mathbb{Z}) \to l^2(\mathbb{Z}), \quad (U_k x)(m) := x(m-k)$$
where $k \in \mathbb{Z}$, and the multiplication operators
$$aI : l^2(\mathbb{Z}) \to l^2(\mathbb{Z}), \quad (ax)(m) := a(m)x(m)$$
where $a \in l^\infty(\mathbb{Z})$. For example, the discrete Schrödinger operator (1.1) is just the band operator $U_{-1} + aI + U_1$. More general, every finite sum $\sum a_k U_k$ with $a_k \in l^\infty(\mathbb{Z})$ is a band operator and, conversely, every band operator can be uniquely written in this way. The functions a_k are called the coefficients of the operator.

Occasionally, we will be interested in band and band-dominated operators, the coefficients of which belong to a certain C^*-subalgebra B of $l^\infty(\mathbb{Z})$. If this algebra is shift-invariant, i.e., if the function $n \mapsto a(n-k)$ belongs to B for every function $a \in B$ and every $k \in \mathbb{Z}$, then the band operators with coefficients in B form an algebra the closure of which we denote by $\mathcal{A}_B(\mathbb{Z})$. In particular, $\mathcal{A}_\mathbb{C}(\mathbb{Z})$ stands for the C^*-algebra of all band-dominated operators with constant coefficients, i.e., for the C^*-algebra of all Laurent operators with continuous generating function. The latter means that for each operator $A \in \mathcal{A}_\mathbb{C}(\mathbb{Z})$, there is a continuous function a on the unit circle $\mathbb{T}$ such that the matrix representation of A with respect to the standard basis is just $(a_{i-j})_{i,j \in \mathbb{Z}}$ where a_j is the jth Fourier coefficient of a,
$$a_j := \frac{1}{2\pi} \int_0^{2\pi} a(e^{it}) e^{-ijt}\, dt.$$

The Laurent operator with generating function a will be denoted by $L(a)$. An overview on Laurent operators as well as on the closely related class of Toeplitz operators with continuous generating functions, including their Fredholmness, invertibility, and stability of the finite sections method, can be found in [**13, 14, 15, 18**], for example. Another remarkable class of coefficients is $SO(\mathbb{Z})$, the algebra of the slowly oscillating functions. By definition, a function $a \in l^\infty(\mathbb{Z})$ is *slowly oscillating* if $a(m+1) - a(m)$ tends to zero as $m \to \pm\infty$. For a closer acquaintance with the Fredholm properties of band-dominated operators with slowly oscillating coefficients one can consult [**32**] and Section 2.4 in [**37**].

Sometimes we will also have to work on l^2-spaces over the non-negative integers $\mathbb{Z}_+ = \mathbb{N}$ and over the negative integers $\mathbb{Z}_-$. We identify $l^2(\mathbb{Z}_+)$ and $l^2(\mathbb{Z}_-)$ with closed subspaces of $l^2(\mathbb{Z})$ in the obvious way, and we denote the orthogonal projections from $l^2(\mathbb{Z})$ onto $l^2(\mathbb{Z}_+)$ and $l^2(\mathbb{Z}_-)$ by P and Q, respectively. Thus, $P + Q = I$ and $PQ = QP = 0$. Band and band-dominated operators over $l^2(\mathbb{Z}_+)$ and $l^2(\mathbb{Z}_-)$ as well as the corresponding algebras $\mathcal{A}(\mathbb{Z}_+) = \mathcal{A}(\mathbb{N})$ and $\mathcal{A}(\mathbb{Z}_-)$ are defined in complete analogy to the case of operators on $l^2(\mathbb{Z})$. In particular, $\mathcal{A}_\mathbb{C}(\mathbb{N})$ is the C^*-subalgebra of $L(l^2(\mathbb{N}))$ which is generated by all Toeplitz operators with continuous generating function, i.e., by the operators
$$T(a) := PL(a)P : l^2(\mathbb{N}) \to l^2(\mathbb{N})$$

with $a \in C(\mathbb{T})$. Similarly, $\mathcal{A}_{SO}(\mathbb{N})$ refers to the C^*-algebra which is generated by the set of all band operators on $l^2(\mathbb{N})$ with slowly oscillating coefficients, where we agree upon calling a bounded function a on $\mathbb{N}$ *slowly oscillating* if
$$\lim_{m \to +\infty} (a(m+1) - a(m)) = 0.$$
The compression PU_kP of the shift operator U_k to $l^2(\mathbb{N})$ will be denoted by V_k.

About this paper. The paper is organized as follows. We start with recalling some basic facts about Fredholmness of band-dominated operators which will be cited without proof. A comprehensive treatment of this topic is in [**37**]; see also the references mentioned there. The basic idea is to express the Fredholmness of a band-dominated operator in terms of uniform invertibility of its limit operators. These results are then applied to study the stability of the finite sections method for band-dominated operators in Section 2. Whereas the material of Sections 2.1 - 2.5 is well known (see [**25, 34, 35, 36, 37, 44, 51**]), those of Sections 2.7 and 2.8 is perhaps new. A completely different approach to study the stability of the finite sections method for band-dominated was proposed in [**20**].

In Section 3 we consider stable regularizations of non-stable sequences. Section 3.1 is mathematical folklore, and the results from Section 3.2 (and much more) can be also found in [**15, 23, 47, 53**] (only the proofs of the splitting property are perhaps new). The application of these results to the finite sections method for band-dominated operators is published here for the first time.

Sections 4 and 5 form certainly the heart of this paper. Here a Fredholm theory for approximation sequences is developed. Some of the mentioned definitions and results can be already found in [**23, 43, 47**], but the present paper seems to be the first place where this material is thoroughly developed. So this part could also serve as an introduction into this subject (for the Fredholm theory in fractal algebras one can also consult [**42**]). Also the characterizations of compact and Fredholm sequences (including the determination of their essential rank and α-number, respectively) in the algebras of the finite sections method is completely new.

Fractal approximation sequences are subject to [**23, 41, 46**]. Sequences in the algebra of the finite sections method do not possess the property of fractality in general, but they own a weaker form of fractality which might be called essential fractality and which is discussed in Section 6. Finally, in Section 7, some further applications of the notions and results introduced and proved above are presented. The main themes there are approximation numbers, determinants, Arveson dichotomy, and essential spectra and pseudospectra.

Acknowledgment. I sincerely thank Mrs. Christiane Herdler and Mrs. Magdalena Tabbert for their kind assistance in preparing the figures in the text and adapting my files to the Memoirs' style. I am also indebted to Albrecht Böttcher and Bernd Silbermann for numerous stimulating discussions and valuable remarks.

Added in proof (December 2006). Since this text was submitted to the editors, some new results appeared which could not be incorporated in the text. I would like to mention some of them briefly.

- The most exciting new result is that the uniform boundedness condition in Theorems 2.7 and 2.8 is redundant for general band-dominated operators. The proof is in [**38**]. It is based on a subsequence version of Theorem 2.7.

- The finite sections method for band-dominated operators with almost periodic coefficients is considered in [**39**]. It is shown there that there exists a distinguished subsequence of the full finite sections sequence $(P_n A P_n)$ which behaves extremely well in the sense that its (fractality, convergence, etc.) properties are close to that of a finite sections sequence of a Toeplitz operator with continuous generating function.
- The results of Section 7.8 on the asymptotic behaviour of determinants of finite sections of band-dominated operators have been largely extended in [**45, 17**].
- There is a new proof of the index Theorem 2.3 for band-dominated operators A which employs the splitting property of the singular values of the finite sections $P_n A P_n$ of A and the formula (5.8) for the α-number of the sequence $(P_n A P_n)$ (whereas the original proof in [**34**] is based on K-theoretic arguments). A publication is in preparation.

CHAPTER 2

Stability

For $n \in \mathbb{N}$, consider the orthogonal projections
$$R_n : l^2(\mathbb{Z}) \to l^2(\mathbb{Z}), \quad (R_n x)(m) := \begin{cases} x(m) & \text{if } -n \leq m < n \\ 0 & \text{otherwise} \end{cases}$$
and
$$P_n : l^2(\mathbb{Z}_+) \to l^2(\mathbb{Z}_+), \quad P_n := P R_n P.$$
The *finite sections method* consists in replacing the operator equation $Au = f$ with $A \in L(l^2(\mathbb{Z}))$ by the sequence of the linear systems

(2.1) $$R_n A R_n u_n = R_n f, \quad n \in \mathbb{N}.$$

Here, the $R_n A R_n$ are viewed of as operators on the Hilbert space im R_n, provided with the norm induced by the norm on $L(l^2(\mathbb{Z}))$. The finite sections method is called *stable* if the matrices $R_n A R_n$ are invertible for sufficiently large n and if the norms of their inverses are uniformly bounded. In this case one also says that the sequence $(R_n A R_n)$ is stable. If the finite sections method for A is stable, then there is an $n_0 \in \mathbb{N}$ such that the equations (2.1) are uniquely solvable for each right-hand side $f \in l^2(\mathbb{Z})$ and for each $n \geq n_0$, and their solutions u_n converge to a solution of the equation $Au = f$ in the norm of $l^2(\mathbb{Z})$. Analogously, one defines the stability of the finite sections method $(P_n A P_n)$ for an operator $A \in L(l^2(\mathbb{N}))$.

It is elementary to check that the finite sections method $(R_n A R_n)$ for the operator $A \in L(l^2(\mathbb{Z}))$ is stable if and only if the block diagonal operator

(2.2) $$\text{diag}\,(R_1 A R_1, R_2 A R_2, R_3 A R_3, \ldots),$$

considered as acting on $l^2(\mathbb{N}) = \mathbb{C}^2 \oplus \mathbb{C}^4 \oplus \ldots$, is a Fredholm operator. Analogously, the finite sections method $(P_n A P_n)$ for an operator $A \in L(l^2(\mathbb{N}))$ is stable if and only if the block diagonal operator

$$\text{diag}\,(P_1 A P_1, P_2 A P_2, P_3 A P_3, \ldots),$$

considered as acting on $l^2(\mathbb{N}) = \mathbb{C}^1 \oplus \mathbb{C}^2 \oplus \ldots$, is Fredholm. Recall in this connection that a bounded linear operator A on a Banach space X is called *Fredholm* if its kernel $\ker A := \{x \in X : Ax = 0\}$ and its cokernel $\operatorname{coker} A := X/\operatorname{im} A$ are finite-dimensional linear spaces, and that in this case the number

$$\operatorname{ind} A := \dim \ker A - \dim \operatorname{coker} A$$

is called the *Fredholm index* of A.

It is evident that for a band operator A, the associated operator (2.2) is a band operator again. Similarly, if A is band-dominated, then (2.2) is a band-dominated operator. Thus, the study of the finite sections method for band-dominated operators becomes part of the Fredholm theory for band-dominated operators. The Fredholm properties of band-dominated operators have been intensively studied in

[**35**] and its successors; see also the recent monograph [**37**]. We briefly recall the facts needed in what follows.

2.1. Fredholm band-dominated operators and their indices

Let $\mathcal{H}$ stand for the set of all sequences $h : \mathbb{N} \to \mathbb{Z}$ which tend to infinity in the sense that, given $C > 0$, there is an n_0 such that $|h(n)| > C$ for all $n \geq n_0$. An operator $A_h \in L(l^2(\mathbb{Z}))$ is called the *limit operator* of $A \in L(l^2(\mathbb{Z}))$ with respect to the sequence $h \in \mathcal{H}$ if $U_{-h(n)} A U_{h(n)}$ tends *-strongly to A_h as $n \to \infty$. Notice that every operator possesses at most one limit operator with respect to a given sequence h. The set $\sigma_{op}(A)$ of all limit operators of a given operator A is called the *operator spectrum* of A.

It is not hard to see that every limit operator of a compact operator is 0 and that every limit operator of a Fredholm operator is invertible. A basic result of [**35**] claims that the operator spectrum of a *band-dominated operator* is rich enough in order to guarantee the reverse implications. Here is a summary of the results from [**35**].

THEOREM 2.1. *Let $A \in L(l^2(\mathbb{Z}))$ be a band-dominated operator. Then*

(a) *every sequence $h \in \mathcal{H}$ possesses a subsequence g such that the limit operator A_g exists.*

(b) *the operator A is compact if and only if $\sigma_{op}(A) = \{0\}$.*

(c) *the operator A is Fredholm if and only if each of its limit operators is invertible and if the norms of their inverses are uniformly bounded.*

(d) *if A is a band operator, then A is Fredholm if and only if each of its limit operators is invertible.*

This result can be extended to band-dominated operators on $l^p(\mathbb{Z}^N)$ with operator-valued coefficients where $1 < p < \infty$ and N is a positive integer; see [**37**] for details. An elegant proof of assertions (a)–(c) which also works for band-dominated operators on other discrete groups than $\mathbb{Z}$ has been recently discovered by Roe [**51**], see also Theorem 2.4 below.

The Fredholm criterion established in Theorem 2.1 takes a particular simple form for band-dominated operators with slowly oscillating coefficients [**32**, **37**].

THEOREM 2.2. *Let $A \in \mathcal{A}_{SO}(\mathbb{Z})$ be a band-dominated operator with slowly oscillating coefficients. Then*

(a) *all limit operators of A belong to $\mathcal{A}_{\mathbb{C}}(\mathbb{Z})$, i.e., they are Laurent operators.*

(b) *the operator A is Fredholm if and only if all of its limit operators are invertible.*

Now we turn over to the Fredholm index. Let A be a band-dominated operator on $l^2(\mathbb{Z})$. Then the operators PAQ and QAP are compact (they are of finite rank if A is a band operator), and so are the operators $A - (PAP + Q)(P + QAQ)$ and $A - (P + QAQ)(PAP + Q)$. Hence, a band-dominated operator A is Fredholm if and only if both $PAP + Q$ and $P + QAQ$ are Fredholm operators, and the Fredholm index of A is equal to the sum of the Fredholm indices of $PAP + Q$ and $P + QAQ$. In this case, we call $\text{ind}_+(A) := \text{ind}(PAP + Q)$ and $\text{ind}_-(A) := \text{ind}(P + QAQ)$ the *plus-index* and the *minus-index* of A. Evidently,

$$(2.3) \qquad \text{ind } A = \text{ind}_+(A) + \text{ind}_-(A)$$

for every Fredholm band-dominated operator A. Notice also that the operator spectrum splits into
$$\sigma_{op}(A) = \sigma_+(A) \cup \sigma_-(A)$$
where $\sigma_+(A)$ and $\sigma_-(A)$ stand for the sets of all limit operators of A which correspond to sequences tending to $+\infty$ and to $-\infty$, respectively.

The index result for Fredholm band-dominated operators established in [34] reads as follows.

THEOREM 2.3. *Let $A \in L(l^2(\mathbb{Z}))$ be a Fredholm band-dominated operator. Then*
(a) *for all $B_+ \in \sigma_+(A)$ and $B_- \in \sigma_-(A)$,*
$$\mathrm{ind}_+(B_+) = \mathrm{ind}_+(A) \quad and \quad \mathrm{ind}_-(B_-) = \mathrm{ind}_-(A).$$
(b) *all operators in $\sigma_+(A)$ have the same plus-index, and all operators in $\sigma_-(A)$ have the same minus-index.*
(c) *for arbitrarily chosen operators $B_+ \in \sigma_+(A)$ and $B_- \in \sigma_-(A)$,*
$$\mathrm{ind}\, A = \mathrm{ind}_+(B_+) + \mathrm{ind}_-(B_-). \tag{2.4}$$

Of course, if A is Fredholm, then the operators in $\sigma_+(A)$ also have the same minus-index, which follows from (b), from the invertibility of all limit operators of A, and from (2.3). It is also evident that (b) and (c) are immediate consequences of assertion (a). The proof of Theorem 2.3 is essentially based upon calculations of the K-groups of the C^*-algebra $\mathcal{A}(\mathbb{Z})$ and of related ideals.

Sometimes it will prove useful to have a symbol calculus for band-dominated operators at our disposal. Several versions of a symbol calculus have been proposed in [37], Sections 1.3.2 and 2.2.3 - 2.2.4. The following variant which is perfectly suited to the Hilbert space setting is taken from Roe [51]. We let $\beta\mathbb{Z}$ stand for the Stone-Čech compactification of $\mathbb{Z}$ and write $\partial\mathbb{Z} := \beta\mathbb{Z} \setminus \mathbb{Z}$ for its boundary. If A is a band-dominated operator, then the *-strong closure of the set $\{U_{-k}AU_k : k \in \mathbb{Z}\}$ in $L(l^2(\mathbb{Z}))$ is *-strongly compact (this is essentially assertion (a) of Theorem 2.1). By the universal property of the Stone-Čech compactification, any map from $\mathbb{Z}$ to a compact Hausdorff space X extends uniquely to a continuous map from $\beta\mathbb{Z}$ to X. In particular, the map
$$\mathbb{Z} \to L(l^2(\mathbb{Z})), \quad k \mapsto U_{-k}AU_k$$
extends uniquely to a *-strongly continuous map
$$\mathrm{smb}\, A : \partial\mathbb{Z} \to L(l^2(\mathbb{Z})). \tag{2.5}$$
We call smb A the *symbol* of A and think of smb A as an element of the C^*-algebra $l^\infty(\partial\mathbb{Z}, L(l^2(\mathbb{Z})))$ of all bounded functions on $\partial\mathbb{Z}$ with values in $L(l^2(\mathbb{Z}))$.

THEOREM 2.4. (a) *For $A \in \mathcal{A}(\mathbb{Z})$, the values of the function smb A are exactly the limit operators of A.*
(b) *The mapping smb is a *-homomorphism from $\mathcal{A}(\mathbb{Z})$ into $l^\infty(\partial\mathbb{Z}, L(l^2(\mathbb{Z})))$ the kernel of which is the ideal $K(l^2(\mathbb{Z}))$ of the compact operators.*

2.2. Algebras of matrix sequences

For every sequence δ of positive integers, let $\mathcal{F}^\delta$ stand for the set of all bounded sequences (A_n) of matrices $A_n \in \mathbb{C}^{\delta(n) \times \delta(n)}$. Provided with the operations
$$(A_n) + (B_n) := (A_n + B_n), \quad (A_n)(B_n) := (A_nB_n), \quad (A_n)^* := (A_n^*)$$

and the supremum norm $\|(A_n)\|_{\mathcal{F}} := \sup_n \|A_n\|$, this set becomes a C^*-algebra with identity element (I_n) where I_n refers to the $\delta(n) \times \delta(n)$ identity matrix. We refer to $\mathcal{F}^\delta$ as an *algebra of matrix sequences* and to δ as its *dimension function*. Thus, the algebra of matrix sequences with constant dimension function $\delta = 1$ is $l^\infty(\mathbb{N})$. The special choice of the dimension function is often not of importance; so we will usually simply write $\mathcal{F}$ in place of $\mathcal{F}^\delta$. But one should mention that the results of Sections 4 and 5 become trivial if the dimension function is bounded.

If R is an orthogonal projection on a Hilbert space with finite rank n, then we choose and fix a basis in $\mathrm{im}\, R$ and identify each operator $A \in L(\mathrm{im}\, R)$ with its matrix representation with respect to the chosen basis. Thus, we identify $L(\mathrm{im}\, R)$ with $\mathbb{C}^{n \times n}$. In particular, if $R = P_n$, then we always choose $e_0, e_1, \ldots, e_{n-1}$ as the basis in $\mathrm{im}\, P_n$. In this way we can consider each sequence $(P_n A P_n)$ of the finite sections method for an operator $A \in L(l^2(\mathbb{N}))$ as an element of the algebra $\mathcal{F}^\delta$ with δ the identity mapping. We denote this algebra by $\mathcal{F}(\mathbb{N})$ in what follows. Similarly, if $R = R_n$, then we agree upon choosing $e_{-n}, e_{-n+1}, \ldots, e_{n-1}$ as the basis in $\mathrm{im}\, R_n$. Thus, the finite sections sequence $(R_n A R_n)$ can be viewed as an element of the algebra $\mathcal{F}^\delta$ with dimension function $\delta(n) := 2n$. We denote this algebra by $\mathcal{F}(\mathbb{Z})$.

We will often have to speak about strongly convergent sequences of matrices, at least in the case when the dimension function is monotonically increasing. Basically, there are two ways to introduce strong convergence in this setting. First, one can identify the matrices A_n in $\mathbb{C}^{\delta(n) \times \delta(n)}$ with operators on $\mathrm{im}\, P_{\delta(n)}$. Then a sequence $(A_n) \in \mathcal{F}$ is said to converge strongly to an operator A if the operators $A_n P_{\delta(n)}$, thought of as acting on $l^2(\mathbb{N})$, converge strongly to A. Second, if δ is a monotonically increasing dimension function of even integers, then one can also identify the matrices A_n in $\mathbb{C}^{\delta(n) \times \delta(n)}$ with operators on $\mathrm{im}\, R_{\delta(n)/2}$. In this case, a sequence $(A_n) \in \mathcal{F}$ is called strongly convergent to an operator A if the operators $A_n R_{\delta(n)/2}$, now considered as acting on $l^2(\mathbb{Z})$, converge strongly to A. For example, the sequence $(D_n)_{n \geq 1}$ of the $2n \times 2n$-matrices

$$D_n := \mathrm{diag}\,(1, 0, 0, \ldots, 0)$$

converges strongly to P_1 if considered in the first way, whereas it converges strongly to 0 if considered in the second way.

Let $\mathcal{F}$ be an algebra of matrix sequences with dimension function δ. The set $\mathcal{G} = \mathcal{G}^\delta$ of all sequences in $\mathcal{F}$ which tend to zero in the norm forms a closed ideal of $\mathcal{F}$. We call $\mathcal{G}$ the ideal of the zero sequences associated with $\mathcal{F}$. The ideals of zero sequences associated with $\mathcal{F}(\mathbb{N})$ and $\mathcal{F}(\mathbb{Z})$ will be denoted by $\mathcal{G}(\mathbb{N})$ and $\mathcal{G}(\mathbb{Z})$, respectively. In particular, the ideal $\mathcal{G}^\delta$ with the constant dimension function $\delta = 1$ is $c_0(\mathbb{N})$.

CONVENTION. Unless otherwise stated, $\mathcal{F} = \mathcal{F}^\delta$ stands for an algebra of matrix sequences with fixed dimension function δ, and $\mathcal{G} = \mathcal{G}^\delta$ stands for its associated ideal of zero sequences in what follows. □

The importance of the quotient algebra $\mathcal{F}/\mathcal{G}$ in numerical analysis rests on the following observation, also known as Kozak's theorem (see, e.g., Theorem 1.15 in [**23**]).

THEOREM 2.5. *A sequence $(A_n) \in \mathcal{F}$ is stable if and only if its coset $(A_n) + \mathcal{G}$ is invertible in $\mathcal{F}/\mathcal{G}$.*

Notice also that the norm of a coset $(A_n) + \mathcal{G}$ in the quotient algebra $\mathcal{F}/\mathcal{G}$ is given by

$$\|(A_n) + \mathcal{G}\|_{\mathcal{F}/\mathcal{G}} = \limsup_{n \to \infty} \|A_n\|. \tag{2.6}$$

2.3. Stability vs. Fredholmness

For another perspective on stability, we associate to each sequence $\mathbf{A} = (A_n) \in \mathcal{F}$ the block diagonal operator

$$\operatorname{Op}(\mathbf{A}) := \operatorname{diag}(A_1, A_2, A_3, \ldots) \tag{2.7}$$

considered as acting on $l^2(\mathbb{N}) = \mathbb{C}^{\delta(1)} \oplus \mathbb{C}^{\delta(2)} \oplus \mathbb{C}^{\delta(3)} \oplus \ldots$. The mapping Op provides us with a $*$-homomorphic embedding of the algebra $\mathcal{F}$ into $L(l^2(\mathbb{N}))$ which allows us to think of $\mathcal{F}$ as a closed $*$-subalgebra of $L(l^2(\mathbb{N}))$. Moreover,

$$\operatorname{Op}(\mathcal{F}) \cap K(l^2(\mathbb{N})) = \operatorname{Op}(\mathcal{G}),$$

as one easily checks. Thus, if $\operatorname{Op}(\mathbf{A})$ is a Fredholm operator for a sequence $\mathbf{A} \in \mathcal{F}$, i.e., if the coset $\operatorname{Op}(\mathbf{A}) + K(l^2(\mathbb{N}))$ is invertible in the quotient algebra $L(l^2(\mathbb{N}))/K(l^2(\mathbb{N}))$, then this coset is invertible in $(\operatorname{Op}(\mathcal{F}) + K(l^2(\mathbb{N})))/K(l^2(\mathbb{N}))$ due to the inverse closedness of C^*-algebras. The isomorphisms

$$(\operatorname{Op}(\mathcal{F}) + K(l^2(\mathbb{N})))/K(l^2(\mathbb{N}))$$
$$\cong \operatorname{Op}(\mathcal{F})/(\operatorname{Op}(\mathcal{F}) \cap K(l^2(\mathbb{N}))) \cong \operatorname{Op}(\mathcal{F})/\operatorname{Op}(\mathcal{G}) \cong \mathcal{F}/\mathcal{G}$$

imply that coset $\mathbf{A} + \mathcal{G}$ is invertible in $\mathcal{F}/\mathcal{G}$. Conversely, if $\mathbf{A} + \mathcal{G} \in \mathcal{F}/\mathcal{G}$ is an invertible coset, then the operator $\operatorname{Op}(\mathbf{A})$ is Fredholm. Thus, invertibility in $\mathcal{F}/\mathcal{G}$ is just the same as invertibility in a subalgebra of the Calkin algebra of $L(l^2(\mathbb{N}))$. Summarizing we get the following.

THEOREM 2.6. *A sequence $\mathbf{A} \in \mathcal{F}$ is stable if and only if the operator $\operatorname{Op}(\mathbf{A})$ is Fredholm on $l^2(\mathbb{N})$.*

2.4. Stability of the finite sections method

The stability criterion in Theorem 2.6 seems to be of less use in general. But if one starts with the sequence $\mathbf{A} = (R_n A R_n)$ of the finite sections method of a band-dominated operator A, then one ends up with a band-dominated operator $\operatorname{Op}(\mathbf{A})$ on $l^2(\mathbb{N})$, and Theorem 2.1 applies to study the Fredholmness of $\operatorname{Op}(\mathbf{A})$. Basically, one has to compute the limit operators of $\operatorname{Op}(\mathbf{A})$, which leads to the following result (which is Theorem 3 in [**36**]).

THEOREM 2.7. *Let $A \in L(l^2(\mathbb{Z}))$ be a band-dominated operator. Then the finite sections method $(R_n A R_n)_{n \geq 1}$ is stable if and only if the operator A, all operators*

$$QA_h Q + P \quad \text{with} \quad A_h \in \sigma_+(A)$$

and all operators

$$PA_h P + Q \quad \text{with} \quad A_h \in \sigma_-(A)$$

are invertible on $l^2(\mathbb{Z})$, and if the norms of their inverses are uniformly bounded. The condition of the uniform boundedness of the inverses is redundant if A is a band operator.

Specifying this theorem to the case of band operators on $l^2(\mathbb{N})$ we get the following, where J refers to the unitary operator
$$l^2(\mathbb{Z}) \to l^2(\mathbb{Z}), \quad (Jx)_m := x_{-m-1},$$
and where we define $\sigma_+(A)$ as $\sigma_+(PAP+Q)$ for each band-dominated operator on $l^2(\mathbb{N})$.

THEOREM 2.8. *Let $A \in L(l^2(\mathbb{N}))$ be a band-dominated operator. Then the finite sections method $(P_n A P_n)_{n \geq 1}$ is stable if and only if the operator A and all operators*
$$JQA_h QJ \quad \text{with} \quad A_h \in \sigma_+(A)$$
are invertible on $l^2(\mathbb{N})$ and if the norms of their inverses are uniformly bounded. The condition of the uniform boundedness of the inverses is redundant if A is a band operator.

We give the proof of Theorem 2.8 only; the proof of Theorem 2.7 follows with obvious modifications.

PROOF. Let $A \in \mathcal{A}(\mathbb{N})$, set $\mathbf{A} := (P_n A P_n)$ and let h be a sequence which tends to infinity and for which the limit operator $\operatorname{Op}(\mathbf{A})_h$ exists. We call numbers of the form $n(n+1)/2$ triangular and distinguish between two cases: Either there is a $k \in \mathbb{Z}$ and a subsequence g of h such that $g(n) + k$ is triangular for all n, or there is a subsequence g of h such that the distance from $g(n)$ to the set of all triangular numbers tends to infinity as $n \to \infty$.

In the latter case, we let Δ_n denote the largest triangular number which is less than $g(n)$. Then $l(n) := g(n) - \Delta_n$ defines a sequence l which tends to infinity, and the limit operator $\operatorname{Op}(\mathbf{A})_h = \operatorname{Op}(\mathbf{A})_g$ of $\operatorname{Op}(\mathbf{A})$ coincides with the limit operator A_l of A.

Let now g be a subsequence of h such that each $g(n) + k$ is triangular for some integer k. Then the sequence l defined by $l(n) := g(n) + k$ tends to infinity, the limit operator of $\operatorname{Op}(\mathbf{A})$ with respect to the sequence l exists, and
$$\operatorname{Op}(\mathbf{A})_l = U_{-k} \operatorname{Op}(\mathbf{A})_g U_k.$$
So it will be sufficient to consider the case when $k = 0$. Define $d(n)$ as the positive integer for which the triangular number $d(n)(d(n)+1)/2$ equals $g(n)$. Then the sequence d tends to infinity, the limit operator of A with respect to d exists, and
$$\operatorname{Op}(\mathbf{A})_h = \operatorname{Op}(\mathbf{A})_g = QA_d Q + PAP.$$
Thus, each limit operator of $\operatorname{Op}(\mathbf{A})$ is either a limit operator of A or it is of the form $U_{-k}(QA_d Q + PAP)U_k$ with an integer k and with a limit operator A_d of A.

Conversely, let A_d be a limit operator of A with respect to the sequence d. If we let
$$d(n) := 2\frac{2h(n)(2h(n)+1)}{2} + h(n) = 2h(n)(h(n)+1),$$
then the limit operator $\operatorname{Op}(\mathbf{A})_h$ exists and is equal to A_d, whereas the choice $h(n) := d(n)(d(n)+1)/2 + k$ with an integer k leads to the limit operator
$$U_{-k}(QA_d Q + PAP)U_k$$
of $\operatorname{Op}(\mathbf{A})$ with respect to h. Thus,
$$\sigma_{op}(\operatorname{Op}(\mathbf{A})) = \sigma_{op}(A) \cup \{U_{-k}(QA_h Q + PAP)U_k : k \in \mathbb{Z}, A_h \in \sigma_{op}(A)\}.$$

This shows that the conditions of the theorem are necessary. They are also sufficient: the invertibility of A implies the invertibility of all limit operators of A, and if both A and $QA_hQ + P$ are invertible then the operator $U_{-k}(QA_hQ + PAP)U_k$ is invertible for each integer k. □

Theorem 2.8 suggests to introduce the *operator spectrum of a sequence* $\mathbf{A} = (A_n) \in \mathcal{F}$ as follows: Consider the shifted operators $U_{-n}A_nU_n$ thought of as acting on $l^2(\mathbb{Z})$. If a certain subsequence $(U_{-h(n)}A_{h(n)}U_{h(n)})_{n \geq 1}$ of the sequence $(U_{-n}A_nU_n)_{n \geq 1}$ possesses a *-strong limit $\mathbf{A}_h$ as $n \to \infty$, then we call $\mathbf{A}_h$ a limit operator of the sequence $\mathbf{A}$, and we denote the set of all limit operators of $\mathbf{A}$ by $\sigma_{op,1}(\mathbf{A})$. The reason for this notation will become evident in Section 2.8. Thus, Theorem 2.8 can be also formulated as follows: If $A \in \mathcal{A}(\mathbb{N})$, then the finite sections method is stable if and only if the operator A is invertible, all operators in $\sigma_{op,1}(\mathbf{A})$ are invertible, and the norms of their inverses are uniformly bounded.

There are generalizations of Theorems 2.7 and 2.8 which will sometimes prove useful and which can be verified in the same vein as their predecessors (whereas it seems to be difficult to derive these results via the approach to the finite sections method for band-dominated operators used in [**36**]). We only mention the result for the finite sections (P_nAP_n).

THEOREM 2.9. *Let $A \in L(l^2(\mathbb{Z}_+))$ be a band-dominated operator, and let $\eta : \mathbb{N} \to \mathbb{N}$ be a strongly monotonically increasing sequence. Then the sequence $(P_{\eta(n)}AP_{\eta(n)})_{n \geq 1}$ is stable if and only if the operator A and all operators JQA_hQJ where A_h is a limit operator of A with respect to a subsequence h of η are invertible on $l^2(\mathbb{Z}_+)$ and if the norms of their inverses are uniformly bounded. The condition of the uniform boundedness of the inverses is redundant if A is a band operator.*

Thus, instead of taking all limit operators of A with respect to monotonically increasing sequences h, one has consider only the limit operators with respect to subsequences of η.

2.5. Band-dominated operators with slowly oscillating coefficients

Theorems 2.7 and 2.8 take a particularly simple form for band-dominated operators with slowly oscillating coefficients. The results in this section go back to [**25**] for band operators and to [**44**] in the general case.

A simple necessary condition for the stability of the finite sections method of the operator A is the invertibility of A (Theorem 1.20 in [**23**]). Our first result says that the invertibility of A is also sufficient for the stability of the finite sections method if A is a band operator with slowly oscillating coefficients on $l^2(\mathbb{N})$.

THEOREM 2.10. *Let $A \in L(l^2(\mathbb{N}))$ be a band operator with slowly oscillating coefficients. Then the finite sections method (P_nAP_n) is stable if and only if the operator A is invertible.*

This result is well known in case of band operators on $l^2(\mathbb{N})$ with constant coefficients, i.e., for Toeplitz operators with continuous generating function (and it still holds for Toeplitz operators with piecewise continuous coefficients which are beyond the scope of this text). It should be mentioned that it is an open question to formulate and prove Theorem 2.10 in the system (= matrix) case.

PROOF. Let $A \in L(l^2(\mathbb{N}))$ be a band-dominated operator with slowly oscillating coefficients. If the finite sections method $(P_n A P_n)$ is stable, then A is invertible. Let, conversely, A be an invertible operator. We identify A with the operator $PAP + Q$ acting on $l^2(\mathbb{Z})$. Clearly, this operator is invertible, too. Hence, all limit operators of $PAP+Q$ are invertible by Theorem 2.1 (c). It is easy to check that the part $\sigma_-(PAP + Q)$ of the operator spectrum of $PAP + Q$ consists of the identity operator only. Let A_h be a limit operator in $\sigma_+(PAP + Q)$. Since the coefficients of A (hence, the coefficients of $PAP + Q$) are slowly oscillating, the operator A_h is shift invariant (Theorem 2.2 (a) and Proposition 30 in [**35**]). From 2.4 and 2.5 in [**14**] we conclude that there is a function $a_h \in L^\infty(\mathbb{T})$ such that $A_h = L(a_h)$. Since A_h is a band-dominated operator, one even has $a_h \in C(\mathbb{T})$. Further, an elementary calculation shows that $JQA_h QJ = PJL(a_h)JP$ is just the Toeplitz operator $T(\widetilde{a_h})$ where $\tilde{a}(t) := a(1/t)$ for a function a on the unit circle.

Since the operator A is invertible, the plus-index of $PAP + Q$ is zero. By Theorem 2.3, the plus- and minus-indices of each limit operator of $PAP + Q$ are zero, too. In particular, the index of $QA_h Q + P$ (which is the minus-index of A_h) is zero. This implies that the index of $JQA_h QJ = T(\widetilde{a_h})$ is zero, whence the invertibility of $T(\widetilde{a_h})$ via Coburn's theorem (see [**18**], Chapter 1, Theorem 3.1, for the case of polynomial generating functions and Theorem 2.38 (b) and Corollary 2.40 in [**14**] for the general case).

Via Theorem 2.8, this settles the assertion of Theorem 2.10 in case A is a band operator. For a general band-dominated operator A, we still have to show the uniform boundedness of the norms of the inverses of the operators $QA_h Q + P$ where the A_h run through $\sigma_+(A)$.

For, we show that the limit operators of A can be labeled by the points of a compact metric space, X, in such a way that the mapping which associates a limit operator A^x of A with each point $x \in X$ becomes norm continuous. The same idea has been used in [**35**] in order to show that the uniform boundedness of the inverses of the limit operators in Theorem 2.8 is redundant in case A is a band-dominated operator with slowly oscillating coefficients (see also Section 2.4.9 in [**37**] and [**7**] for further occurrences of the same idea).

Let (A_n) be a sequence of band operators with slowly oscillating coefficients which tends to A in the norm of $L(l^2(\mathbb{Z}_+))$. By SO_A we denote the smallest closed and symmetric subalgebra of $l^\infty(\mathbb{Z}_+)$ which contains all coefficients of all operators A_n as well as all sequences tending to zero. Then SO_A is a commutative and separable (since generated by a countable number of functions) C^*-algebra. We denote its maximal ideal space by $M(SO_A)$ and write $M^{+\infty}(SO_A)$ for its fiber over $+\infty$, i.e., for the collection of all non-trivial multiplicative functionals $f : SO_A \to \mathbb{C}$ which vanish over $c_0(\mathbb{Z}_+)$. Since SO_A is a separable algebra, its maximal ideal space is metrizable, and so is the fiber $X := M^{+\infty}(SO_A)$.

In [**35**] it is shown that if $B = \sum_\alpha b_\alpha P U_\alpha P$ is a band operator with coefficients b_α in SO_A, and if $h \in \mathcal{H}$ is a sequence which tends to $+\infty$ and for which the limit operator B_h exists, then there is a unique point $x \in M^{+\infty}(SO_A)$ such that

$$(2.8) \qquad B_h = \sum b_\alpha(x) U_\alpha =: B^x.$$

Conversely, given $x \in M^{+\infty}(SO_A)$, there is a sequence $h \in \mathcal{H}$ such that the limit operator of B with respect to h exists and that (2.8) holds (the metrizability of $M^{+\infty}(SO_A)$ is used in order to approximate x by a *sequence* of points in $\mathbb{Z}_+$ from

which one can choose h as a subsequence). Evidently, the function
$$M^{+\infty}(SO_A) \to L(l^2(\mathbb{Z})), \quad x \mapsto \sum b_\alpha(x) U_\alpha$$
is continuous.

Now we extend this construction to the operator A. Let h be a sequence which tends to $+\infty$ for which the limit operator A_h exists. By a Cantor diagonal argument, there is a subsequence g of h such that all limit operators $(A_n)_g$ exist, too. Let x denote the uniquely determined point in $M^{+\infty}(SO_A)$ such that $(A_n)_g = A_n^x$. Then
$$A_h = \lim_{n \to \infty} (A_n)_g = \lim_{n \to \infty} A_n^x,$$
and we denote this operator by A^x. Being the uniform limit of the continuous functions $x \mapsto A_n^x$, the function
$$M^{+\infty}(SO_A) \to L(l^2(\mathbb{Z})), \quad x \mapsto A^x$$
is continuous. Then the function
$$M^{+\infty}(SO_A) \to L(l^2(\mathbb{Z})), \quad x \mapsto QA^x Q + P$$
is continuous. Since the values of this function are invertible operators as we have already seen, and since inversion is continuous, the function

(2.9) $$M^{+\infty}(SO_A) \to L(l^2(\mathbb{Z})), \quad x \mapsto \|(QA^x Q + P)^{-1}\|$$

proves to be continuous, too. Finally, being a continuous function on a compact metric space, the function (2.9) attains its maximum. Since the values of that function are exactly the norms of the operators $(QA_h Q + P)^{-1}$ with $A_h \in \sigma_+(A)$, we obtain the uniform boundedness of these norms. $\square$

2.6. The role of the index

One cannot expect that Theorem 2.10 remains valid for band operators with slowly oscillating coefficients on $l^2(\mathbb{Z})$ – it is already wrong for band Laurent operators. In fact, the finite sections method $(R_n L(a) R_n)$ for the *Laurent* operator $L(a)$ is stable if and only if the *Toeplitz* operator $T(a)$ is invertible. (Notice that the invertibility of $T(a)$ implies that of $L(a)$, but if $a(t) := t^{-1}$, then the Laurent operator $L(a)$ is invertible, whereas the Toeplitz operator $T(a)$ has a non-trivial kernel.) The additional condition which is needed besides the invertibility of A in order to guarantee the stability of the finite sections method $(R_n A R_n)$ can be expressed in terms of the plus-index of A.

PROPOSITION 2.11. *Let $A \in L(l^2(\mathbb{Z}))$ be a band-dominated operator for which the finite sections method $(R_n A R_n)$ is stable. Then A is invertible, and the plus-index of A is zero.*

PROOF. The stability of $(R_n A R_n)$ implies the invertibility of A for each operator $A \in L(l^2(\mathbb{Z}))$. Let now A be band-dominated, and let A_h be a limit operator of A which lies in $\sigma_+(A)$. Then $QA_h Q + P$ is invertible by Theorem 2.7, whence
$$0 = \operatorname{ind}(QA_h Q + P) = \operatorname{ind}_-(A_h) = \operatorname{ind}_-(A)$$
by Theorem 2.3. Since A is invertible, this implies that $\operatorname{ind}_+(A) = 0$. $\square$

In case A is a band-dominated operator with slowly oscillating coefficients, these necessary conditions prove to be sufficient.

THEOREM 2.12. *Let $A \in L(l^2(\mathbb{Z}))$ be a band-dominated operator with slowly oscillating coefficients. Then the finite sections method $(R_n A R_n)$ is stable if and only if the operator A is invertible and if the plus-index of A is zero.*

If these conditions are satisfied, then, by (2.3), the minus-index of A is zero, too. Notice also that for operators with constant coefficients, i.e., for band-dominated Laurent operators $A = L(a)$, the Fredholmness of $L(a)$ implies the Fredholmness of the Toeplitz operator $T(a)$ and that the plus-index of A is just the Fredholm index of $T(a)$. Since Fredholm Toeplitz operators with index zero are invertible (Coburn's theorem), we rediscover the classical result for the finite sections method for band-dominated Laurent operators from Theorem 2.12.

PROOF. The *only if* part of the assertion follows from the preceding proposition. For the *if* part, let A be invertible and $\mathrm{ind}_+(A) = 0$, and let A_h be a limit operator in $\sigma_\pm(A)$. Then also $\mathrm{ind}_-(A) = 0$, and we get as above that
$$0 = \mathrm{ind}_\pm(A) = \mathrm{ind}_\pm(A_h),$$
whence
$$\mathrm{ind}\,(P A_h P + Q) = \mathrm{ind}\,(Q A_h Q + P) = 0.$$
The shift invariance of the limit operators of A implies that the operators $P A_h P + Q$ and $Q A_h Q + P$ can be identified with Toeplitz operators. Since the indices of these operators are zero, Coburn's theorem implies the invertibility of these operators. The uniform boundedness of the inverses of the operators $P A_h P + Q$ and $Q A_h Q + P$ can be verified as in the proof of Theorem 2.10 (where one now has to work with both fibers $M^\pm(SO_A)$). □

Theorems 2.10 and 2.12 remain valid for band-dominated operators acting on l^p-spaces with $1 < p < \infty$, see [**44**].

2.7. The C^*-algebra of the finite sections method

The goal of this section is to examine the smallest closed subalgebra $\mathcal{S}(\mathbb{N})$ of $\mathcal{F} = \mathcal{F}(\mathbb{N})$ which contains all sequences $(P_n A P_n)$ with a band-dominated operator $A \in \mathcal{A}(\mathbb{N})$. We call $\mathcal{S}(\mathbb{N})$ the *algebra of the finite sections method*. We will point out that the structure of the algebra $\mathcal{S}(\mathbb{N})$ is surprisingly similar to the structure of the algebra $\mathcal{S}_{\mathbb{C}}(\mathbb{N})$ of the finite sections method for band-dominated Toeplitz operators, i.e., the smallest closed subalgebra of $\mathcal{F}$ which contains all sequences $(P_n T(a) P_n)$ with $a \in C(\mathbb{T})$. The following theorem summarizes briefly what is known about the structure of the latter algebra (see, for example, Theorem 1.53 in [**23**]). For $n \in \mathbb{N}$, we will need the reflection operators
$$W_n : l^2(\mathbb{N}) \to l^2(\mathbb{N}), \quad (x_0, x_1, \ldots) \mapsto (x_{n-1}, x_{n-2}, \ldots, x_0, 0, 0, \ldots).$$
Notice that these operators converge weakly to zero; hence, the operators $W_n L W_n$ converge strongly to zero for each compact operator L.

THEOREM 2.13. *(a) Every sequence $(A_n) \in \mathcal{S}_{\mathbb{C}}(\mathbb{N})$ can be uniquely written as*
$$(2.10) \quad (A_n) = (P_n T(a) P_n) + (P_n K P_n) + (W_n L W_n) + (G_n)$$
where $a \in C(\mathbb{T})$, the operators K and L are compact, and $(G_n) \in \mathcal{G} = \mathcal{G}(\mathbb{N})$.
(b) The set
$$(2.11) \quad \{(W_n L W_n) + (G_n) : L \text{ compact}, (G_n) \in \mathcal{G}\}$$

coincides with the quasicommutator ideal of the algebra $\mathcal{S}_\mathbb{C}(\mathbb{N})$, i.e., with the smallest closed ideal of this algebra which contains all sequences
$$(P_n T(a) P_n)(P_n T(b) P_n) - (P_n T(a) T(b) P_n)$$
with $a, b \in C(\mathbb{T})$.

Observe also that the sequences in the quasicommutator ideal (2.11) are localized at the ends of the interval $\{0, 1, \ldots, n-1\}$ in a sense which will be made precise in assertion (d) of Theorem 2.15 below. To illustrate this point, let K and L be given by the $r \times r$-matrices (k_{ij}) and (l_{ij}) with respect to the standard basis of $l^2(\mathbb{N})$. Then, for $n > 2r$, the matrix representation of $P_n K P_n + W_n L W_n$ is

$$\begin{pmatrix} k_{00} & \cdots & k_{0,r-1} & 0 & \cdots & 0 & 0 & \cdots & 0 \\ \vdots & & \vdots & \vdots & & \vdots & \vdots & & \vdots \\ k_{r-1,0} & \cdots & k_{r-1,r-1} & 0 & \cdots & 0 & 0 & \cdots & 0 \\ 0 & \cdots & 0 & 0 & \cdots & 0 & 0 & \cdots & 0 \\ \vdots & & \vdots & \vdots & \ddots & \vdots & \vdots & & \vdots \\ 0 & \cdots & 0 & 0 & \cdots & 0 & 0 & \cdots & 0 \\ 0 & \cdots & 0 & 0 & \cdots & 0 & l_{r-1,r-1} & \cdots & l_{r-1,0} \\ \vdots & & \vdots & \vdots & & \vdots & \vdots & & \vdots \\ 0 & \cdots & 0 & 0 & \cdots & 0 & l_{0,r-1} & \cdots & l_{00} \end{pmatrix}.$$

For the generalization of Theorem 2.13 to the finite sections algebra for general band-dominated operators, let $\mathcal{K}_0(\mathbb{N})$ refer to the set of all sequences in $\mathcal{S}(\mathbb{N})$ which converge *-strongly to zero. Clearly, $\mathcal{K}_0(\mathbb{N})$ is a closed ideal of the algebra $\mathcal{S}(\mathbb{N})$, and the following becomes evident.

THEOREM 2.14. *Every sequence* $(A_n) \in \mathcal{S}(\mathbb{N})$ *can be uniquely written as*
(2.12) $$(A_n) = (P_n A P_n) + (K_n)$$
with an operator $A \in \mathcal{A}(\mathbb{N})$ *and a sequence* $(K_n) \in \mathcal{K}_0(\mathbb{N})$.

Sequences in $\mathcal{K}_0(\mathbb{N})$ can be characterized as follows.

THEOREM 2.15. *The following conditions are equivalent for a sequence* $(K_n) \in \mathcal{S}(\mathbb{N})$:
(a) (K_n) *tends strongly to zero;*
(b) (K_n) *belongs to the quasicommutator ideal of the algebra* $\mathcal{S}(\mathbb{N})$, *i.e., to the smallest closed ideal of* $\mathcal{S}(\mathbb{N})$ *which contains all sequences*
$$(P_n A P_n)(P_n B P_n) - (P_n A B P_n)$$
with band operators A and B;
(c) (K_n) *is localized at the right-hand end of the interval* $\{0, 1, \ldots, n-1\}$ *in the sense that, given* $\varepsilon > 0$, *there are non-negative integers* n_0 *and* l_0 *such that*
(2.13) $$\sup_{n \geq n_0} \|P_{n-l_0} K_n P_{n-l_0}\| < \varepsilon;$$
(d) (K_n) *is localized at the right-hand end of the interval* $\{0, 1, \ldots, n-1\}$ *in the sense that, given* $\varepsilon > 0$, *there are non-negative integers* n_0 *and* l_0 *such that*
(2.14) $$\sup_{n \geq n_0} \|K_n - (P_n - P_{n-l_0}) K_n (P_n - P_{n-l_0})\| < \varepsilon.$$

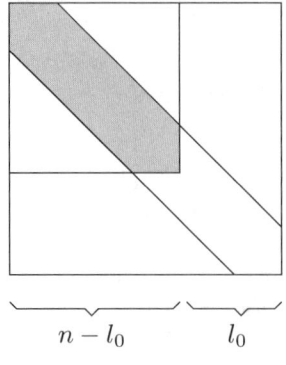
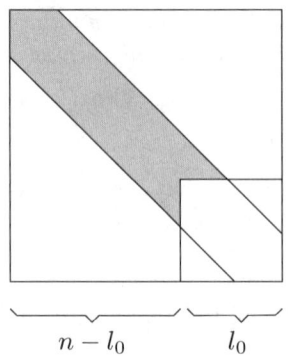

$$\underbrace{}_{n-l_0}\underbrace{}_{l_0} \qquad \underbrace{}_{n-l_0}\underbrace{}_{l_0}$$

$$P_{n-l_0} K_n P_{n-l_0} \qquad\qquad K_n - (P_n - P_{n-l_0}) K_n (P_n - P_{n-l_0})$$

For example, this theorem implies that the algebra $\mathcal{S}(\mathbb{N})$ cannot contain sequences of the form (C_n) with $C_n = \operatorname{diag}(0, \ldots, 0, 1, 0, \ldots, 0)$ with the 1 standing at the $[n/2]$th position or of the form (D_n) with $D_n = \operatorname{diag}(1/n, 2/n, \ldots, n/n)$. Indeed, the sequence (C_n) is localized at the middle of the interval $\{0, 1, \ldots, n-1\}$, whereas the sequence (D_n) is not localized at all.

PROOF. We denote the quasicommutator ideal of the algebra $\mathcal{S}(\mathbb{N})$ by $\mathcal{K}_1$, the set of all sequences in $\mathcal{S}(\mathbb{N})$ subject to condition (c) by $\mathcal{K}_2$, and the set of all sequences in $\mathcal{S}(\mathbb{N})$ subject to condition (d) by $\mathcal{K}_3$. We start with the inclusion $\mathcal{K}_0(\mathbb{N}) \subseteq \mathcal{K}_1$. Let $(K_n) \in \mathcal{S}(\mathbb{N})$ be a sequence with strong limit zero. Given $\varepsilon > 0$, choose band operators A_{ij} such that

$$\|(K_n) - \sum_i \prod_j (P_n A_{ij} P_n)\|_{\mathcal{F}} < \varepsilon/2.$$

Further abbreviate $\sum_i \prod_j A_{ij}$ to A. Letting n go to infinity in this inequality, one gets $\|A\| < \varepsilon/2$, and via induction one easily obtains that

$$(J_n) := \sum \prod (P_n A_{ij} P_n) - (P_n A P_n)$$

belongs to the quasicommutator ideal $\mathcal{K}_1$. Hence,

$$\begin{aligned}\|(K_n) - (J_n)\| &\leq \|(K_n) - \sum_i \prod_j (P_n A_{ij} P_n)\| + \|(P_n A P_n)\| \\ &\leq \|(K_n) - \sum_i \prod_j (P_n A_{ij} P_n)\| + \|A\| < \varepsilon,\end{aligned}$$

showing that (K_n) can be approximated as closely as desired by sequences in $\mathcal{K}_1$. Since $\mathcal{K}_1$ is closed by its definition, $\mathcal{K}_0(\mathbb{N}) \subseteq \mathcal{K}_1$.

To verify the inclusion $\mathcal{K}_1 \subseteq \mathcal{K}_2$, we first show that $\mathcal{K}_2$ is a closed ideal of $\mathcal{S}(\mathbb{N})$. Let (K_n^0), (K_n^1) be sequences in $\mathcal{K}_2$. Given $\varepsilon > 0$, choose n_0, n_1 and l_0, l_1 such that

$$\sup_{n \geq n_0} \|P_{n-l_0} K_n^0 P_{n-l_0}\| < \varepsilon/2$$

and

$$\sup_{n \geq n_1} \|P_{n-l_1} K_n^1 P_{n-l_1}\| < \varepsilon/2.$$

Set $n^* := \max\{n_0, n_1\}$ and $l^* := \max\{l_0, l_1\}$. Then, for all $n \geq n^*$,

$$\|P_{n-l^*}(K_n^0 + K_n^1)P_{n-l^*}\|$$
$$\leq \|P_{n-l^*}K_n^0 P_{n-l^*}\| + \|P_{n-l^*}K_n^1 P_{n-l^*}\|$$
$$\leq \|P_{n-l_0}K_n^0 P_{n-l_0}\| + \|P_{n-l_1}K_n^1 P_{n-l_1}\| < \varepsilon$$

since $P_{n-l^*}P_{n-l_i} = P_{n-l_i}$ for $i = 0, 1$. Hence, $(K_n^0 + K_n^1) \in \mathcal{K}_2$.

Let now (K_n) be a sequence in $\mathcal{K}_2$ and A a non-zero band operator. Given $\varepsilon > 0$, choose n_0, l_0 such that

$$\sup_{n \geq n_0} \|P_{n-l_0}K_n P_{n-l_0}\| < \varepsilon/\|A\|,$$

and choose n_1, l_1 such that

$$P_n A P_n P_{n-l_1} = P_{n-l_0} P_n A P_n P_{n-l_1} \quad \text{for all } n \geq n_1$$

which is possible since A is a band operator. Set $n^* := \max\{n_0, n_1\}$ and $l^* := \max\{l_0, l_1\}$. Then, for all $n \geq n^*$,

$$\|P_{n-l^*}K_n(P_n A P_n)P_{n-l^*}\|$$
$$\leq \|P_{n-l^*}P_{n-l_0}K_n(P_n A P_n)P_{n-l_1}P_{n-l^*}\|$$
$$\leq \|P_{n-l_0}K_n P_{n-l_0}(P_n A P_n)P_{n-l_1}\|$$
$$\leq \|P_{n-l_0}K_n P_{n-l_0}\|\|A\| \leq \varepsilon.$$

Hence, $(K_n)(P_n A P_n) \in \mathcal{K}_2$, and analogously, $(P_n A P_n)(K_n) \in \mathcal{K}_2$.

Finally we show that $\mathcal{K}_2$ is closed in $\mathcal{F}$. Let the sequence $(K_n) \in \mathcal{F}$ be the limit (in the norm of $\mathcal{F}$) of a sequence of sequences in $\mathcal{K}_2$. Thus, given $\varepsilon > 0$, there is a sequence $(K_n^0) \in \mathcal{K}_2$ with $\|(K_n) - (K_n^0)\|_{\mathcal{F}} < \varepsilon/2$. If n_0 and l_0 are chosen such that

$$\sup_{n \geq n_0} \|P_{n-l_0}K_n^0 P_{n-l_0}\| < \varepsilon/\|A\|,$$

then, for all $n \geq n_0$,

$$\|P_{n-l_0}K_n P_{n-l_0}\|$$
$$\leq \|P_{n-l_0}K_n^0 P_{n-l_0}\| + \|P_{n-l_0}(K_n - K_n^0)P_{n-l_0}\|$$
$$\leq \|P_{n-l_0}K_n^0 P_{n-l_0}\| + \|(K_n) - (K_n^0)\|_{\mathcal{F}} < \varepsilon.$$

Thus, $(K_n) \in \mathcal{K}_2$, and $\mathcal{K}_2$ is indeed a closed ideal of $\mathcal{S}(\mathbb{N})$. Now the inclusion $\mathcal{K}_1 \subseteq \mathcal{K}_2$ can be seen as follows. Let A, B be band operators. Due to the band property, there are integers n_0, l_0 and n_1, l_1 such that

$$P_{n-l_0}A Q_n = 0 \quad \text{and} \quad Q_n B P_{n-l_1} = 0$$

for all $n \geq n_0$ and $n \geq n_1$, respectively. For $n \geq n^* := \max\{n_0, n_1\}$ and $l^* := \max\{l_0, l_1\}$ one thus obtains

$$P_{n-l^*}P_n A Q_n B P_n P_{n-l^*} = P_{n-l^*}A Q_n B P_{n-l^*} = 0.$$

Thus, the sequence

$$(P_n A B P_n) - (P_n A P_n)(P_n B P_n) = (P_n A Q_n B P_n)$$

belongs to $\mathcal{K}_2$. Since these sequences generate the quasicommutator ideal, one gets $\mathcal{K}_1 \subseteq \mathcal{K}_2$.

For the inclusion $\mathcal{K}_2 \subseteq \mathcal{K}_3$, let $(K_n) \in \mathcal{K}_2$. Given $\varepsilon > 0$, choose n_0, l_0 such that

$$\sup_{n \geq n_0} \|P_{n-l_0}K_n P_{n-l_0}\| < \varepsilon.$$

Let further $(K_n^0) \in \mathcal{S}(\mathbb{N})$ be a sequence of matrices with finite band width l_1 independent of n such that $\|(K_n) - (K_n^0)\|_{\mathcal{F}} < \varepsilon$. From

$$\begin{aligned}&K_n^0 - (P_n - P_{n-l_0-l_1})K_n^0(P_n - P_{n-l_0-l_1})\\ &= P_{n-l_0}K_n^0 P_{n-l_0} - (P_{n-l_0} - P_{n-l_0-l_1})K_n^0(P_{n-l_0} - P_{n-l_0-l_1})\end{aligned}$$

we conclude

$$\begin{aligned}&\|K_n^0 - (P_n - P_{n-l_0-l_1})K_n^0(P_n - P_{n-l_0-l_1})\|\\ &\leq \|P_{n-l_0}K_n^0 P_{n-l_0}\| + \|(P_{n-l_0} - P_{n-l_0-l_1})K_n^0(P_{n-l_0} - P_{n-l_0-l_1})\|\\ &\leq 2\|P_{n-l_0}K_n^0 P_{n-l_0}\|\\ &\leq 2\|P_{n-l_0}K_n P_{n-l_0}\| + 2\|(K_n) - (K_n^0)\|_{\mathcal{F}} \leq 4\varepsilon,\end{aligned}$$

whence

$$\begin{aligned}&\|K_n - (P_n - P_{n-l_0-l_1})K_n(P_n - P_{n-l_0-l_1})\|\\ &\leq \|K_n^0 - (P_n - P_{n-l_0-l_1})K_n^0(P_n - P_{n-l_0-l_1})\| + 2\|(K_n) - (K_n^0)\|_{\mathcal{F}} \leq 6\varepsilon.\end{aligned}$$

Thus, $(K_n) \in \mathcal{K}_3$. For the final inclusion $\mathcal{K}_3 \subseteq \mathcal{K}_0(\mathbb{N})$, let $(K_n) \in \mathcal{K}_3$, and let $x \in l^2(\mathbb{N})$ be a finitely supported sequence, say $x(n) = 0$ for $n \geq n_0$. Given $\varepsilon > 0$, choose n_1, l_1 such that

$$\sup_{n \geq n_1} \|K_n - (P_n - P_{n-l_1})K_n(P_n - P_{n-l_1})\| < \varepsilon.$$

Then, for $n \geq \max\{n_0 + l_1, n_1\}$,

$$\|K_n x\| = \|K_n P_{n_0} x\| = \|(K_n - (P_n - P_{n-l_1})K_n(P_n - P_{n-l_1}))P_{n_0}x\| \leq \varepsilon \|x\|,$$

whence the assertion. $\square$

To establish the analogous result for the finite sections method $(R_n A R_n)$ for a band-dominated operator A on $l^2(\mathbb{Z})$, recall that $\mathcal{F}(\mathbb{Z})$ refers to the C^*-algebra of all bounded sequences (A_n) of $2n \times 2n$-matrices A_n and $\mathcal{G}(\mathbb{Z})$ to the ideal of all sequences in $\mathcal{F}(\mathbb{Z})$ which tend to zero. By means of the identification of $\mathbb{C}^{2n \times 2n}$ with $L(\operatorname{im} R_n)$ we can think of the finite sections sequences $(R_n A R_n)$ as elements of the algebra $\mathcal{F}(\mathbb{Z})$. Let $\mathcal{S}(\mathbb{Z})$ denote the smallest closed subalgebra of $\mathcal{F}(\mathbb{Z})$ which contains all finite sections sequences $(R_n A R_n)$ with $A \in \mathcal{A}(\mathbb{Z})$, and write $\mathcal{K}_0(\mathbb{Z})$ for the closed ideal of $\mathcal{S}(\mathbb{Z})$ which consists of all sequences which converge strongly to zero. Then the analogs of Theorems 2.14 and 2.15 read as follows.

THEOREM 2.16. *Every sequence $(A_n) \in \mathcal{S}(\mathbb{Z})$ can be uniquely written as*

(2.15) $$(A_n) = (R_n A R_n) + (K_n)$$

with an operator $A \in \mathcal{A}(\mathbb{Z})$ and a sequence $(K_n) \in \mathcal{K}_0(\mathbb{Z})$.

THEOREM 2.17. *The following conditions are equivalent for a sequence $(K_n) \in \mathcal{S}(\mathbb{Z})$:*
(a) *(K_n) tends strongly to zero;*
(b) *(K_n) belongs to the quasicommutator ideal of the algebra $\mathcal{S}(\mathbb{Z})$, i.e., to the smallest closed ideal of $\mathcal{S}(\mathbb{Z})$ which contains all sequences*

$$(R_n A R_n)(R_n B R_n) - (R_n A B R_n)$$

with band operators A and B;

(c) (K_n) is localized at the ends of the interval $\{-n, -n+1, \ldots, n-1\}$ in the sense that, given $\varepsilon > 0$, there are non-negative integers n_0 and l_0 such that

(2.16)
$$\sup_{n \geq n_0} \|R_{n-l_0} K_n R_{n-l_0}\| < \varepsilon;$$

(d) (K_n) is localized at the ends of the interval $\{-n, -n+1, \ldots, n-1\}$ in the sense that, given $\varepsilon > 0$, there are non-negative integers n_0 and l_0 such that

(2.17)
$$\sup_{n \geq n_0} \|K_n - (R_n - R_{n-l_0}) K_n (R_n - R_{n-l_0})\| < \varepsilon.$$

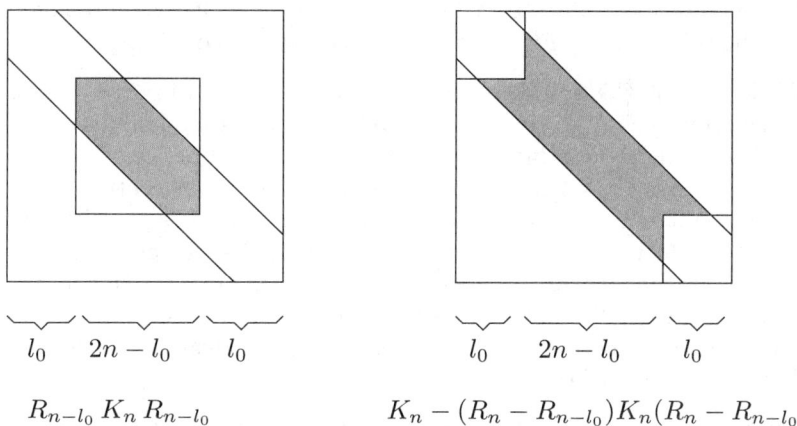

$\underbrace{\qquad}_{l_0} \underbrace{\qquad}_{2n-l_0} \underbrace{\qquad}_{l_0} \qquad \underbrace{\qquad}_{l_0} \underbrace{\qquad}_{2n-l_0} \underbrace{\qquad}_{l_0}$

$R_{n-l_0} K_n R_{n-l_0} \qquad\qquad K_n - (R_n - R_{n-l_0}) K_n (R_n - R_{n-l_0})$

The proofs are close to those of Theorems 2.14 and 2.15 and we omit the details.

2.8. Another perspective on localization

Here we discuss another interpretation of what is meant by saying that a sequence (K_n) is localized at the right-hand end of the interval $\{0, 1, \ldots, n-1\}$.

Let $L^2(\mathbb{R})$ stand for the Hilbert space of all squared integrable functions over the real line and, for each positive integer n, let S_n denote its closed subspace spanned by the functions

$$\varphi_{kn}(t) := \sqrt{n}\chi_{[0,\,1)}(nt - k) \qquad \text{with } k \in \mathbb{Z}.$$

Let L_n refer to the orthogonal projection of $L^2(\mathbb{R})$ onto S_n. These projections converge strongly to the identity operator on $L^2(\mathbb{R})$ as n tends to infinity. Another basic fact known in the spline space community as the *commutator property of spline projections* states that, for every bounded and uniformly continuous function f on $\mathbb{R}$,

(2.18)
$$\lim_{n \to \infty} \|L_n fI - f L_n\|_{L(L^2(\mathbb{R}))} = 0$$

(see Theorem 2.7 in [**22**]).

We let $\mathcal{F}(\mathbb{R})$ stand for the set of all bounded sequences (A_n) of bounded linear operators A_n on S_n. Provided with elementwise operations and with the supremum norm

$$\|(A_n)\|_{\mathcal{F}(\mathbb{R})} := \sup \|A_n L_n\|_{L(L^2(\mathbb{R}))},$$

the set $\mathcal{F}(\mathbb{R})$ becomes a C^*-algebra, and its subset $\mathcal{G}(\mathbb{R})$ which is formed by all sequences tending to zero in the norm is a closed ideal of $\mathcal{F}(\mathbb{R})$. The C^*-algebra of all bounded and uniformly continuous functions on $\mathbb{R}$ will be denoted by BUC.

PROPOSITION 2.18. *The set of all cosets $(L_n f|_{S_n}) + \mathcal{G}(\mathbb{R})$ with $f \in BUC$ is a C^*-subalgebra of $\mathcal{F}(\mathbb{R})/\mathcal{G}(\mathbb{R})$ which is *-isomorphic to BUC.*

PROOF. The commutator property implies that

(2.19) $$(L_n f|_{S_n})(L_n g|_{S_n}) - (L_n fg|_{S_n}) \in \mathcal{G}(\mathbb{R})$$

for all functions $f, g \in BUC$. This proves the first assertion. For the second assertion notice that the mapping

$$BUC \to \mathcal{F}(\mathbb{R})/\mathcal{G}(\mathbb{R}), \quad f \mapsto (L_n f|_{S_n}) + \mathcal{G}(\mathbb{R})$$

is a *-homomorphism due to (2.19), the kernel of which is zero since the strong limit of the operators $L_n f L_n$ as $n \to \infty$ exists and is equal to f. $\square$

We let $\mathcal{F}(\mathbb{R})^{comm}$ stand for the set of all sequences in $\mathcal{F}(\mathbb{R})$ which commute with each sequence $(L_n f|_{S_n})$ with f a BUC functions modulo a sequence in $\mathcal{G}(\mathbb{R})$. Thus, BUC can be identified with a central subalgebra of $\mathcal{F}(\mathbb{R})/\mathcal{G}(\mathbb{R})$, and the latter algebra can be localized over BUC in the sense of central localization. To state this result which is also known as the *local principle by Allan and Douglas*, let $\mathcal{B}$ be a C^*-algebra with identity, and let $\mathcal{C}$ be a C^*-subalgebra of the center of $\mathcal{B}$ which contains the identity. Being central means that every element of $\mathcal{C}$ commutes with every element of $\mathcal{B}$. Thus, $\mathcal{C}$ is a commutative C^*-algebra. Its maximal ideal space will be denoted by $M(\mathcal{C})$. For every maximal ideal x of $\mathcal{C}$, let I_x denote the smallest closed ideal of $\mathcal{B}$ which contains x.

THEOREM 2.19 (Allan/Douglas). *Let $\mathcal{B}$, $\mathcal{C}$, $M(\mathcal{C})$ and I_x be as above. Then I_x is a proper ideal of $\mathcal{B}$ for every $x \in M(\mathcal{C})$, and the following assertions are equivalent for $b \in \mathcal{B}$:*

(a) b is invertible in $\mathcal{B}$.

(b) $a + I_x$ is invertible in the quotient algebra $\mathcal{B}/I_x$ for each $x \in M(\mathcal{C})$.

Proofs can be found in [**1, 16**] and in [**22**], Section 1.4.4. Let us see how this localization can be used for band-dominated operators. For, we have to embed the algebra of the band-dominated operators into the algebra $\mathcal{F}(\mathbb{R})$. This embedding is provided by means of the mappings

$$E_n : l^2(\mathbb{Z}) \to S_n, \quad x \mapsto \sum_{k \in \mathbb{Z}} x(m) \varphi_{kn}$$

and

$$E_{-n} : S_n \to l^2(\mathbb{Z}), \quad \sum_{k \in \mathbb{Z}} x(m) \varphi_{kn} \mapsto x$$

which are surjective isometries satisfying $E_n^{-1} = E_{-n}$.

PROPOSITION 2.20. *If A is band-dominated, then the sequence $(E_n A E_{-n})$ belongs to $\mathcal{F}(\mathbb{R})^{comm}$.*

PROOF. Since the mapping $A \mapsto (E_n A E_{-n})$ is a *-homomorphism from $\mathcal{A}(\mathbb{Z})$ into $\mathcal{F}(\mathbb{R})$, we need to prove the assertion only for the generating operators U, U^{-1} and aI with $a \in l^\infty(\mathbb{Z})$ of $\mathcal{A}(\mathbb{Z})$ in place of A.

Let $A = U$. One easily checks that, for every positive integer n and for every function $f \in BUC$,

$$E_n U E_{-n} L_n f I|_{S_n} = L_n f_n E_n U E_{-n}$$

where $f_n(x) = f(x - \frac{1}{n})$. Hence,

$$\|E_n U E_{-n} L_n f I|_{S_n} - L_n f E_n U E_{-n}\| = \|L_n(f_n - f) E_n U E_{-n}\|$$
$$\leq \|f_n - f\|_\infty,$$

and the right-hand side of this estimate tends to zero as $n \to \infty$ since f is uniformly continuous. This shows that the sequence $(E_n U E_{-n})$ is in $\mathcal{F}(\mathbb{R})^{comm}$, and the proof for the sequence $(E_n U^{-1} E_{-n})$ proceeds analogously.

Let now $A = aI$ with $a \in l^\infty(\mathbb{Z})$, and let $f \in BUC$ again. Then $E_n A E_{-n}$ is the operator of multiplication by the function

$$\varphi_n^A(x) := \frac{1}{\sqrt{n}} \sum_{k \in \mathbb{Z}} a_k \varphi_{kn}(x)$$

acting on S_n, and the commutator property implies

$$\|E_n A E_{-n} L_n f I|_{S_n} - L_n f E_n A E_{-n}\|$$
$$= \|\varphi_n^A L_n f I|_{S_n} - L_n f \varphi_n^A I|_{S_n}\|$$
$$= \|\varphi_n^A (L_n f I - f L_n) I|_{S_n} - (L_n f I - f L_n) \varphi_n^A I|_{S_n}$$
$$\quad + \varphi_n^A f L_n|_{S_n} - f L_n \varphi_n^A I|_{S_n}\|$$
$$\leq 2\|\varphi_n^A\|_\infty \|L_n f I - f L_n\| + \|\varphi_n^A f L_n|_{S_n} - f L_n \varphi_n^A I|_{S_n}\|$$
$$= 2\|\varphi_n^A\|_\infty \|L_n f I - f L_n\| \to 0$$

as $n \to \infty$ (notice that the operator of multiplication by φ_n^A maps S_n into S_n). Thus, $(E_n A E_{-n}) \in \mathcal{F}(\mathbb{R})^{comm}$. □

Since $E_n P_n E_{-n}$ is just the operator of multiplication by the characteristic function of the interval $[0, 1]$, it becomes evident that being localized at the right-hand end of the interval $\{0, 1, \ldots, n-1\}$ in the sense of Theorem 2.15 is the same as being localized at the right-hand end of the interval $[0, 1]$ in the sense of central localization. To make this precise, we let I_x denote the smallest closed ideal of the algebra $\mathcal{F}_\mathbb{R}^{comm}/\mathcal{G}(\mathbb{R})$ which contains the maximal ideal x of BUC, and we write Φ_x for the canonical homomorphism from $\mathcal{F}_\mathbb{R}^{comm}$ onto $(\mathcal{F}_\mathbb{R}^{comm}/\mathcal{G}(\mathbb{R}))/I_x$. Observe also that each point $x \in \mathbb{R}$ gives rise to a maximal ideal of BUC via evaluation at x. We denote this ideal by x again.

THEOREM 2.21. *If $(K_n) \in \mathcal{K}_0(\mathbb{N})$, then $\Phi_x(E_n K_n E_{-n}) = 0$ for each $x \neq 1$ in $M(BUC)$.*

Thus, the sequence $(E_n K_n E_{-n})$ is localized at the point 1.

PROOF. Let $x \in M(BUC) \setminus \{1\}$. We choose an open neighborhood $U = (1-\delta, 1+\delta)$ of 1 with some $\delta > 0$ and a continuous function $f \in BUC$ such that $f(x) = 1$ and $f|_U = 0$. Then

(2.20) $$\Phi_x(E_n K_n E_{-n}) = \Phi_x(L_n f E_n K_n E_{-n} L_n f I|_{S_n})$$

since $\Phi_x(L_n f I|_{S_n})$ is the identity element in $(\mathcal{F}_\mathbb{R}^{comm}/\mathcal{G}(\mathbb{R}))/I_x$. The operator $F_n := E_{-n} L_n f E_n \in L(l^2(\mathbb{Z}))$ is a diagonal operator the entries $f_j^{(n)}$ of which are zero if

$[n(1-\delta)] + 1 < j < [n(1+\delta)] - 1$. Thus,
$$\begin{aligned}\|L_n f E_n K_n E_{-n} L_n f I|_{S_n}\| &= \|E_{-n} L_n f E_n K_n E_{-n} L_n f E_n\| \\ &= \|F_n P_n K_n P_n F_n\| \\ &= \|F_n P_{[n(1-\delta)]+1} K_n P_{[n(1-\delta)]+1} F_n\| \\ &\leq \|f\|^2 \|P_{[n(1-\delta)]+1} K_n P_{[n(1-\delta)]+1}\|.\end{aligned}$$

Let now $\varepsilon > 0$. Then, by Theorem 2.15 (c), there are non-negative integers n_0 and l_0 such that
$$\sup_{n \geq n_0} \|P_{n-l_0} K_n P_{n-l_0}\| < \varepsilon.$$
Choose $N_0 \geq n_0$ such that $[n(1-\delta)] + 1 < n - l_0$ for all $n \geq N_0$. For these n,
$$\begin{aligned}\|P_{[n(1-\delta)]+1} K_n P_{[n(1-\delta)]+1}\| &= \|P_{[n(1-\delta)]+1} P_{n-l_0} K_n P_{n-l_0} P_{[n(1-\delta)]+1}\| \\ &\leq \|P_{n-l_0} K_n P_{n-l_0}\| < \varepsilon.\end{aligned}$$

Thus, $(L_n f E_n K_n E_{-n} L_n f I|_{S_n})$ is a sequence in $\mathcal{G}(\mathbb{R})$, whence $\Phi_x(E_n K_n E_{-n}) = 0$ via (2.20). $\square$

Analogously, one has $\Phi_x(E_n K_n E_{-n}) = 0$ for each sequence $(K_n) \in \mathcal{K}_0(\mathbb{Z})$ and each $x \in M(BUC) \setminus \{-1, 1\}$.

CHAPTER 3

Stable regularizability

Let $\mathcal{F}$ be an algebra of matrix sequences, and let $(A_n) \in \mathcal{F}$ be a sequence which converges strongly to an operator A. If infinitely many of the matrices A_n are not invertible then one might come to the idea to consider the Moore-Penrose inverses $A_n^\dagger$ of A_n in place of the common inverses. The resulting questions concerning the *-strong convergence of the sequence $(A_n^\dagger)$ are quite intriguing and not yet completely understood even in case (A_n) is the finite sections method for a Toeplitz operator. Moore-Penrose inverses also raise serious problems from the computational point of view since the computation of least square solutions is extremely sensitive with respect to small perturbations. Thus, even in case one is able to prove convergence, any perturbation to (A_n) be it ever so small can disturb this convergence completely.

In 1974, Moore and Nashed [26] proposed to determine certain approximations of the Moore-Penrose inverses of the A_n rather than the Moore-Penrose inverses themselves. Their idea led to the notion of a stably regularizable sequence introduced and studied for the finite sections method for Toeplitz operators in [53].

DEFINITION 3.1. A sequence $(A_n) \in \mathcal{F}$ is *stably regularizable* if there exists a sequence $(B_n) \in \mathcal{F}$ such that

$$\|A_n B_n A_n - A_n\| \to 0, \quad \|B_n A_n B_n - B_n\| \to 0,$$
$$\|(A_n B_n)^* - A_n B_n\| \to 0, \quad \|(B_n A_n)^* - B_n A_n\| \to 0.$$

Thus, a sequence $(A_n) \in \mathcal{F}$ is stably regularizable if its coset $(A_n) + \mathcal{G}$ is Moore-Penrose invertible in the quotient algebra $\mathcal{F}/\mathcal{G}$.

3.1. Moore-Penrose invertibility in C^*-algebras

We prepare the study of stably regularizable sequences by recalling some basic facts on Moore-Penrose invertibility in C^*-algebras.

Let $\mathcal{B}$ be a C^*-algebra. An element $a \in \mathcal{B}$ is said to be *Moore-Penrose invertible* if there is a $b \in \mathcal{B}$ such that

(3.1) $\qquad aba = a, \quad bab = b, \quad (ab)^* = ab \quad \text{and} \quad (ba)^* = ba.$

If b_1 and b_2 are elements of $\mathcal{B}$ satisfying (3.1) in place of b, then

$$b_1 = b_1 a b_1 = b_1 (a b_1)^* = b_1 b_1^* a^* = b_1 b_1^* (a^* b_2^* a^*)$$
$$= b_1 (b_1^* a^*)(b_2^* a^*) = b_1 (a b_1)^* (a b_2)^* = b_1 (a b_1 a) b_2 = b_1 a b_2$$

and, similarly, $b_2 = b_1 a b_2$, whence $b_1 = b_2$. Thus, the element b in (3.1) is unique (if it exists). It is called the *Moore-Penrose inverse* of a, and we denote it by $a^\dagger$. One easily checks that a and a^* are Moore-Penrose invertible only simultaneously and that $(a^*)^\dagger = (a^\dagger)^*$ and $(a^*a)^\dagger = a^\dagger (a^*)^\dagger$, whereas the identity $(ab)^\dagger = b^\dagger a^\dagger$

is wrong in general. Notice also that $aa^\dagger$ and $a^\dagger a$ are projections, i.e., self-adjoint idempotents.

The following theorem summarizes some equivalent conditions for the Moore-Penrose invertibility in unital C^*-algebras. These facts are mathematical folklore. The spectrum of an element c of a unital C^*-algebra is denoted by $\sigma(c)$.

THEOREM 3.2. *Let $\mathcal{B}$ be a C^*-algebra with identity e. The following conditions are equivalent for an element a of $\mathcal{B}$:*

(a) The element a is Moore-Penrose invertible.

(b) The element a^*a is invertible, or 0 is an isolated point of $\sigma(a^*a)$.

(c) There is a projection p in $\mathrm{alg}\,(e, a^*a)$ ($=$ the smallest closed subalgebra of $\mathcal{B}$ which contains e and a^*a) such that $ap = 0$ and $a^*a + p$ is invertible.

If one of these conditions is satisfied, then the projection p is uniquely determined, and $a^\dagger = (a^*a + p)^{-1}a^*$. Moreover,

$$\|a^\dagger\|^2 = \sup\{1/\lambda : \lambda \in \sigma(a^*a) \setminus \{0\}\}.$$

PROOF. $(a) \Rightarrow (b)$: Let a be Moore-Penrose invertible and set $b := a^\dagger$. If $a = 0$, then 0 is an isolated point of $\sigma(a^*a)$. If $a \neq 0$, then $b \neq 0$, and $e - \lambda bb^*$ is invertible in $\mathcal{B}$ for every complex number λ with $0 < |\lambda| < \|bb^*\|^{-1}$ by Neumann series. Straightforward calculation shows that $(e - \lambda bb^*)^{-1}bb^* - 1/\lambda(e - ba)$ is the inverse of $a^*a - \lambda e$, i.e., either $0 \notin \sigma(a^*a)$, or 0 is an isolated point of that spectrum.

$(b) \Rightarrow (c)$: If a^*a is invertible, one can choose $p = 0$. So let 0 be an isolated point of $\sigma(a^*a)$. By the Gelfand-Naimark theorem, the commutative C^*-algebra $\mathrm{alg}\,(e, a^*a)$ $*$-isomorphic to the C^*-algebra $C(X)$ where $X = \sigma(a^*a)$. We denote the Gelfand transform of an element $c \in \mathrm{alg}\,(e, a^*a)$ by c again. Set $X_0 := \{x \in X : (a^*a)(x) = 0\}$ and $X_1 := X \setminus X_0$. Assumption (b) guarantees that both sets X_0 and X_1 are open and closed subsets of X. Hence,

$$x \mapsto \begin{cases} 0 & \text{if } x \in X_1 \\ 1 & \text{if } x \in X_0 \end{cases}$$

defines a continuous function on X. Let p denote the (uniquely determined) element of $\mathrm{alg}\,(e, a^*a)$ which has this function as its Gelfand transform. One easily checks (by considering Gelfand transforms again) that $a^*ap = 0$ and that $a^*a + p$ is invertible. Since then $pa^*ap = 0$ and $\|pa^*ap\| = \|(ap)^*(ap)\| = \|ap\|^2$, we get assertion (c).

$(c) \Rightarrow (a)$: It is straightforward to check that $(a^*a + p)^{-1}a^*$ is the Moore-Penrose inverse of a. Furthermore, from $ap = 0$ one gets $(a^*a + p)p = p$, whence

(3.2) $$(a^*a + p)^{-1}p = p,$$

whereas the identity $a^\dagger = (a^*a + p)^{-1}a^*$ implies

(3.3) $$(a^*a + p)^{-1}a^*a = a^\dagger a.$$

Addition of (3.2) and (3.3) gives $p = e - a^\dagger a$, showing the uniqueness of the projection p. Finally, the norm identity follows from

$$\|a^\dagger\|^2 = \|(a^\dagger)^*\|^2 = \|a^\dagger(a^\dagger)^*\| = \|a^\dagger(a^*)^\dagger\| = \|(a^*a)^\dagger\|$$

by employing the isometry of the Gelfand transform. $\square$

The (uniquely determined) projection p in (c) is called the *Moore-Penrose projection* of a.

One peculiarity of C^*-algebras is their inverse closedness with respect to common invertibility. It is an immediate consequence of Theorem 3.2 that C^*-algebras are also inverse closed with respect to Moore-Penrose invertibility.

COROLLARY 3.3. *Let $\mathcal{B}$ be a C^*-algebra with identity and $\mathcal{C}$ a C^*-subalgebra of $\mathcal{B}$ containing the identity. If $c \in \mathcal{C}$ is Moore-Penrose invertible in $\mathcal{B}$, then $c^\dagger \in \mathcal{C}$.*

Indeed, the inverse closedness with respect to the common invertibility implies $\sigma_\mathcal{B}(c^*c) = \sigma_\mathcal{C}(c^*c)$, and the equivalence (a) $\Leftrightarrow$ (b) in Theorem 3.2 yields the assertion. □

COROLLARY 3.4. *Let $\mathcal{B}$ be a C^*-algebra (without identity) and $\widetilde{\mathcal{B}}$ its minimal unitization. If $b \in \mathcal{B}$ is Moore-Penrose invertible in $\widetilde{\mathcal{B}}$, then $b^\dagger \in \mathcal{B}$.*

Since $\mathcal{B}$ is an ideal of $\widetilde{\mathcal{B}}$, this remark follows immediately from the identity $a^\dagger = (a^*a + p)^{-1}a^*$. □

EXAMPLE. Let $\mathcal{B} = L(H)$ be the C^*-algebra of all bounded linear operators on a Hilbert space H. An operator $A \in L(H)$ is Moore-Penrose invertible if and only if its range is closed in H. In this case, the Moore-Penrose projection is just the orthogonal projection of H onto the kernel of A. In particular, every bounded linear operator acting on a finite-dimensional Hilbert space is Moore-Penrose invertible (see, e.g., Example 2.16 in [**23**]). The operators with closed range are also called *normally solvable*. □

3.2. Stable regularizability vs. Moore-Penrose invertibility

From Theorem 2.6 we know that the stability of a sequence $\mathbf{A} \in \mathcal{F}$ corresponds to the Fredholmness of the operator Op $(\mathbf{A})$, i.e., to the invertibility of the coset Op $(\mathbf{A}) + K(l^2(\mathbb{N}))$ in the Calkin algebra of $l^2(\mathbb{N})$. Here is the analogous result for stable regularizability.

THEOREM 3.5. *The following conditions are equivalent for a sequence $\mathbf{A} \in \mathcal{F}$:*
(a) the sequence $\mathbf{A}$ is stably regularizable.
(b) the coset $\mathbf{A} + \mathcal{G}$ is Moore-Penrose invertible in $\mathcal{F}/\mathcal{G}$.
(c) the coset Op $(\mathbf{A}) + K(l^2(\mathbb{N}))$ is Moore-Penrose invertible in the Calkin algebra $L(l^2(\mathbb{N}))/K(l^2(\mathbb{N}))$.

The equivalence of (a) with (b) is evident, and that of (a) and (c) can be shown as in the proof of Theorem 2.6 due to the inverse closedness of C^*-algebras with respect to Moore-Penrose invertibility by Corollary 3.3. □

We will now employ Theorem 3.5 to derive an equivalent description of the stable regularizability of a sequence (A_n) in terms of the asymptotic behaviour of the singular values of the matrices A_n. This phenomenon is known as the *splitting property* of the singular values. The singular values of an $n \times n$ matrix A are the non-negative square roots of the eigenvalues of A^*A. We denote them by

$$0 \leq \sigma_1(A) \leq \sigma_2(A) \leq \ldots \leq \sigma_n(A) = \|A\|$$

and write $\sigma_{sing}(A)$ for the set of all singular values of A.

THEOREM 3.6. *The sequence $(A_n) \in \mathcal{F}$ is stably regularizable if and only if there are numbers $d > 0$ and $c_n \geq 0$ with $\lim_{n \to \infty} c_n = 0$ such that*

(3.4) $$\sigma_{sing}(A_n) \subset [0, c_n] \cup [d, \infty) \quad \text{for all } n \in \mathbb{N}.$$

Thus, the set $\sigma_{sing}(A_n)$ splits asymptotically into two parts: one which contracts to 0 as n increases, and one which is separated from 0 by a positive constant for all n.

PROOF. Set $\mathbf{A} := (A_n)$ and $B := \text{Op}(\mathbf{A})$. The sequence (A_n) is stably regularizable if and only if the coset $B + K(l^2(\mathbb{N}))$ is Moore-Penrose invertible in the Calkin algebra. By Theorem 3.2, the latter is equivalent to the fact that the essential spectrum $\sigma_{ess}(B^*B)$ of B^*B (i.e., the spectrum of the coset $B^*B + K(l^2(\mathbb{N}))$ in the Calkin algebra) is contained in $\{0\} \cup [d, \infty)$ for some positive constant d. We choose d to be the largest positive number with this property. From the Fredholm theory of bounded self-adjoint operators it is further known that the intersection

$$(\sigma(B^*B) \setminus \sigma_{ess}(B^*B)) \cap (0, d)$$

consists of an at most countable number of points which are eigenvalues of B^*B of finite multiplicity, and which can accumulate only at 0 or d (Corollary XI.8.5 in [**19**]).

Let now (A_n) be a stably regularizable sequence and assume contrary to what we want to show that there is an interval $[c', d'] \subset (0, d)$ such that the intersection $\sigma(A_n^* A_n) \cap [c', d']$ is non-empty for infinitely many matrices A_n. Since each eigenvalue of $A_n^* A_n$ is also an eigenvalue of B^*B, this would give us infinitely many eigenvalues of B^*B in $[c', d']$ which contradicts the above cited statement. Hence, $\sigma(A_n^* A_n) \cap [c', d']$ is empty for all sufficiently large n, which implies the splitting property (3.4).

Let, conversely, the splitting property (3.4) hold for a sequence $(A_n) \in \mathcal{F}$. Then the sequence $(A_n^* A_n) - \lambda(I_n)$ is stable for each $\lambda \in (0, d)$ as one easily checks via the Gelfand-Naimark theorem for commutative C^*-algebras. Thus, $\sigma(\mathbf{A}^*\mathbf{A} + \mathcal{G}) \cap (0, d) = \emptyset$ whence the Moore-Penrose invertibility of $\mathbf{A} + \mathcal{G}$. □

A direct proof of Theorem 3.6 (without referring to the relation between stable regularizability and Fredholmness and to the relation between the spectrum and the essential spectrum of a bounded linear operator) has been given in [**47**]; see also Theorems 2.14 and 2.19 in [**23**]. A third (and perhaps the most simple) proof of the splitting property follows by employing the identity

$$\limsup_{n \to \infty} \sigma(A_n^* A_n) = \sigma(\mathbf{A}^* \mathbf{A} + \mathcal{G})$$

holding for all sequences $\mathbf{A} = (A_n) \in \mathcal{F}$ (which is Corollary 3.18 in [**23**]).

3.3. Stable regularizability of the finite sections method

If A is a band-dominated operator on $l^2(\mathbb{N})$ and $\mathbf{A} = (P_n A P_n)$, then $\text{Op}(\mathbf{A})$ is a band-dominated operator on $l^2(\mathbb{N})$, and Theorem 3.5 reduces the problem of stable regularizability of $\mathbf{A}$ to the problem of the Moore-Penrose invertibility in the Calkin algebra of a certain band-dominated operator.

THEOREM 3.7. *Let $A \in \mathcal{A}(\mathbb{Z})$. Then the coset $A + K(l^2(\mathbb{Z}))$ is Moore-Penrose invertible in the Calkin algebra $L(l^2(\mathbb{Z}))/K(l^2(\mathbb{Z}))$ if and only if all limit operators* of A are Moore-Penrose invertible and if the norms of their Moore-Penrose inverses are uniformly bounded.

PROOF. Let A be a band-dominated operator with Moore-Penrose invertible coset $A + K(l^2(\mathbb{Z}))$. By inverse closedness of C^*-algebras with respect to Moore-Penrose invertibility, the Moore-Penrose inverse of $A + K(l^2(\mathbb{Z}))$ belongs to the quotient algebra $\mathcal{A}(\mathbb{Z})/K(l^2(\mathbb{Z}))$. Thus, there are a band-dominated operator B and compact operators K_i such that

(3.5) $\quad ABA = A + K_1, \, BAB = B + K_2, \, (AB)^* = AB + K_3, \, (BA)^* = BA + K_4.$

If now $h \in \mathcal{H}$ is a sequence such that the limit operator A_h exists, then there is a subsequence g of h such that the limit operators B_g and $(K_i)_g$ exist, too (Theorem 2.1 (a)). Passing to limit operators with respect to g in (3.5) yields the Moore-Penrose invertibility of $A_h = A_g$, and it shows moreover that $B_g = (A_g)^\dagger$ whence $\|(A_g)^\dagger\| \le \|B\|$.

Let, conversely, all limit operators of A be uniformly Moore-Penrose invertible. The symbol of A defined by (2.5) is Moore-Penrose invertible in $l^\infty(\partial \mathbb{Z}, L(l^2(\mathbb{Z})))$. By inverse closedness with respect to Moore-Penrose invertibility again, smb A is Moore-Penrose invertible in the C^*-subalgebra smb $\mathcal{A}(\mathbb{Z})$ of $l^\infty(\partial \mathbb{Z}, L(l^2(\mathbb{Z})))$. Since the kernel of the $*$-homomorphism smb consists exactly of the compact operators, the coset $A + K(l^2(\mathbb{Z}))$ is Moore-Penrose invertible in the Calkin algebra. □

For completeness, we still consider the special case of band-dominated operators with slowly oscillating coefficients.

THEOREM 3.8. (a) Let $A \in \mathcal{A}_{SO}(\mathbb{N})$. Then the coset $A + K(l^2(\mathbb{N}))$ is Moore-Penrose invertible in the Calkin algebra $L(l^2(\mathbb{N}))/K(l^2(\mathbb{N}))$ if and only if either A is compact or A is Fredholm.

(b) Let $A \in \mathcal{A}_{SO}(\mathbb{Z})$. Then the coset $A + K(l^2(\mathbb{Z}))$ is Moore-Penrose invertible in the Calkin algebra $L(l^2(\mathbb{Z}))/K(l^2(\mathbb{Z}))$ if and only if either A is compact, or PAP is compact and $P + QAQ$ is Fredholm, or QAQ is compact and $PAP + Q$ is Fredholm, or A is Fredholm.

PROOF. We will need a few more notations. Let $M(SO(\mathbb{N}))$ denote the maximal ideal space of the commutative C^*-algebra $SO(\mathbb{N})$, and by $M^\infty(SO(\mathbb{N})) = M(SO(\mathbb{N})) \setminus \mathbb{N}$ denote its fiber at infinity. This fiber is connected as has been shown in Theorem 2.4.7 in [37]. Given a band operator $B = \sum a_k V_k$ on $l^2(\mathbb{N})$ with slowly oscillating coefficients a_k, we associate to B the function

(3.6) $\quad \operatorname{Smb} B : M^\infty(SO(\mathbb{N})) \times \mathbb{T} \to \mathbb{C}, \quad (\tau, \xi) \mapsto \sum \widehat{a_k}(\tau) \xi^k$

where $\widehat{a}$ refers to the Gelfand transform of $a \in SO(\mathbb{N})$. It has been shown in Section 2.4.9 of [37] that the mapping Smb extends to a $*$-homomorphism from $\mathcal{A}_{SO}(\mathbb{N})$ to the algebra of all continuous functions on $M^\infty(SO(\mathbb{N})) \times \mathbb{T}$, and that the kernel of this mapping is the ideal of the compact operators.

Now we turn over to the proof of assertion (a). Clearly, both the compactness and the Fredholmness of A imply the Moore-Penrose invertibility of the coset $A + K(l^2(\mathbb{N}))$. Let, conversely, this coset be Moore-Penrose invertible in the Calkin algebra. Then the symbol of A defined by (3.6) is a Moore-Penrose invertible continuous function on $M^\infty(SO(\mathbb{N})) \times \mathbb{T}$. Since the latter space is connected, there are no non-trivial projections in $C(M^\infty(SO(\mathbb{N})) \times \mathbb{T})$. Thus, either Smb A is the zero function (in which case A is compact) or $\operatorname{Smb} A$ is invertible (in which case A is Fredholm).

The proof of assertion (b) proceeds similarly. Now one has to take into account that both fibers $M^{+\infty}(SO(\mathbb{Z}))$ and $M^{-\infty}(SO(\mathbb{Z}))$ of $M(SO(\mathbb{Z}))$ at $+\infty$ and at $-\infty$ are connected (which is shown in Theorem 2.4.7 in [**37**] again). □

A combination of Theorems 3.5 and 3.7 with Theorems 2.8 and 2.7 gives the following characterizations of the stable regularizability of the finite sections method for band-dominated operators in terms of their limit operators.

THEOREM 3.9. *Let $A \in L(l^2(\mathbb{N}))$ be a band-dominated operator. Then the finite sections method $(P_n A P_n)$ is stably regularizable if and only if the operator A and all operators*

$$JQA_hQJ \quad \text{with} \quad A_h \in \sigma_+(A)$$

are Moore-Penrose invertible on $l^2(\mathbb{N})$ and if the norms of their Moore-Penrose inverses are uniformly bounded.

THEOREM 3.10. *Let $A \in L(l^2(\mathbb{Z}))$ be a band-dominated operator. Then the finite sections method $(R_n A R_n)$ is stably regularizable if and only if the operator A, all operators*

$$QA_hQ + P \quad \text{with} \quad A_h \in \sigma_+(A)$$

and all operators

$$PA_hP + Q \quad \text{with} \quad A_h \in \sigma_-(A)$$

are Moore-Penrose invertible on $l^2(\mathbb{Z})$, and if the norms of their Moore-Penrose inverses are uniformly bounded.

For the proof one has to make use of the fact that $PAP + QAQ$ is Moore-Penrose invertible if and only if PAP and QAQ are Moore-Penrose invertible. It is an open question whether the condition of the *uniform* boundedness of the Moore-Penrose inverses in the preceding two theorems becomes redundant if A is a band operator.

For band-dominated operators with slowly oscillating coefficients, one can say more. First notice that an operator $A \in \mathcal{A}_{SO}(\mathbb{N})$ is Moore-Penrose invertible if and only if it is Fredholm or of finite rank. Indeed, if A is Moore-Penrose invertible, then it is Moore-Penrose invertible in the Calkin algebra. Hence, it is Fredholm or compact by Theorem 3.8 (a). Since Fredholm operators have closed range, they are Moore-Penrose invertible, whereas a compact operator is Moore-Penrose invertible if and only if it has finite rank.

THEOREM 3.11. *Let $A \in \mathcal{A}_{SO}(\mathbb{N})$ be a Fredholm operator. Then the finite sections method $(P_n A P_n)$ is stably regularizable.*

PROOF. If A is Fredholm then each limit operator A_h of A is invertible. From the proof of Theorem 2.10 we recall that then each operator JQA_hQJ is a Fredholm Toeplitz operator, hence Moore-Penrose invertible. By Theorem 3.9, it remains to show that the norms of the corresponding Moore-Penrose inverses are uniformly bounded.

As in the proof of Theorem 2.10, we introduce the algebra SO_A and obtain a norm-continuous function

(3.7) $$M^{+\infty}(SO_A) \to L(l^2(\mathbb{N})), \quad x \mapsto JQA^xQJ$$

the values of which are the operators of the form JQA_hQJ with A_h running through the operator spectrum of A. As mentioned above, each of the values of the function (3.7) is Moore-Penrose invertible. We claim that the function

$$(3.8) \qquad M^{+\infty}(SO_A) \to L(l^2(\mathbb{N})), \quad x \mapsto (JQA^xQJ)^\dagger$$

is norm-continuous, too. Notice that, in contrast to common invertibility, Moore-Penrose inversion is not a continuous operation. But is *is* continuous when restricted to the class $\Phi_{k,l}$ of all Fredholm operators on $l^2(\mathbb{N})$ of fixed kernel dimension k and fixed cokernel dimension l ([**21**]; see also [**48**] for a proof which is close to the spirit of the present paper). This result applies in the context of the function (3.7). Indeed, if (T_n) is a sequence of Fredholm Toeplitz operators which converges in the norm to a Fredholm Toeplitz operator T, then the indices $\operatorname{ind} T_n$ and $\operatorname{ind} T$ coincide for large n due to the continuity of the index. Then also the kernel and cokernel dimensions of T_n coincide with those of T due to Coburn's theorem.

Thus, the function (3.8) is continuous, and the assertion follows as in the proof of Theorem 2.10. □

An analogous result holds for operators in $\mathcal{A}_{SO}(\mathbb{Z})$.

CHAPTER 4

Compactness

The goal of this and the following section is to develop a theory of compact and Fredholm approximation sequences which is parallel to the theory of compact and Fredholm operators. Of particular interest will be the α-number of a Fredholm sequence which corresponds to the kernel dimension of a Fredholm operator. The α-numbers of the finite sections sequences of Toeplitz operators have found fruitful applications in the numerical determination of kernel dimensions of Toeplitz operators and of partial indices of matrix functions [43, 55, 54].

In the following sections, let $\mathcal{F}$ be an algebra of matrix sequences with a strongly monotonically increasing dimension function δ. Thus, $\delta(n) \geq n$.

4.1. Compact sequences

A sequence (K_n) in the C^*-algebra $\mathcal{F}$ is a *sequence of rank one matrices* if every matrix K_n has range dimension less than or equal to one. The smallest closed ideal of $\mathcal{F}$ which contains all sequences of rank one matrices will be denoted by $\mathcal{K}$. The product of a sequence of rank one matrices with another sequence in $\mathcal{F}$ is again a sequence of rank one matrices. Hence, the set $\mathcal{K}_0$ of all finite sums of sequences of rank one matrices forms an (in general, non-closed) ideal of $\mathcal{F}$ the closure of which is just the ideal $\mathcal{K}$. Consequently, a sequence $(A_n) \in \mathcal{F}$ belongs to $\mathcal{K}$ if and only if, for every $\varepsilon > 0$, there is a sequence $(K_n) \in \mathcal{F}$ such that

(4.1) $$\sup_n \|A_n - K_n\| < \varepsilon \quad \text{and} \quad \sup_n \operatorname{rank} K_n < \infty.$$

We refer to the elements of $\mathcal{K}$ as *compact sequences*. The role of the ideal $\mathcal{K}$ in numerical analysis can be compared with the role of the ideal of the compact operators in operator theory.

Notice that $\mathcal{K}$ encloses the ideal $\mathcal{G}$. Indeed, given a sequence $(A_n) \in \mathcal{G}$ and an $\varepsilon > 0$, set $K_n := A_n$ if $\|A_n\| \geq \varepsilon$ and $K_n := 0$ otherwise. Then (4.1) is satisfied since there are only finitely many operators K_n which are not zero.

An appropriate notion of the rank of a sequence in $\mathcal{F}$ can be introduced as follows. We say that a sequence $\mathbf{A} \in \mathcal{F}$ has *finite essential rank* if it is the sum of a sequence (G_n) in $\mathcal{G}$ and of a sequence (K_n) with $\sup_n \operatorname{rank} K_n < \infty$. If $\mathbf{A}$ is of finite essential rank, then there is a smallest integer $r \geq 0$ such that $\mathbf{A}$ can be written as $(G_n) + (K_n)$ with $(G_n) \in \mathcal{G}$ and $\sup_n \operatorname{rank} K_n \leq r$. We call this integer the *essential rank* of $\mathbf{A}$ and write

$$\operatorname{ess\,rank} \mathbf{A} = r.$$

If $\mathbf{A}$ is not of finite essential rank, then we put $\operatorname{ess\,rank} \mathbf{A} = \infty$. Thus, the sequences of essential rank 0 are just the sequences in $\mathcal{G}$. Clearly, the sequences of finite

essential rank form an ideal of $\mathcal{F}$ which is dense in $\mathcal{K}$, and
$$\text{ess rank}\,(\mathbf{A}+\mathbf{B}) \leq \text{ess rank}\,\mathbf{A} + \text{ess rank}\,\mathbf{B},$$
$$\text{ess rank}\,(\mathbf{AB}) \leq \min\{\text{ess rank}\,\mathbf{A},\,\text{ess rank}\,\mathbf{B}\}$$
for arbitrary sequences $\mathbf{A}, \mathbf{B} \in \mathcal{F}$.

Let us emphasize that, in general, the spectral points of a compact sequence do not show the behavior one might expect from the behavior of the eigenvalues of a compact operator. Indeed, if K is a compact operator on a Hilbert space, then there are at most countably many eigenvalues, and all non-zero eigenvalues are isolated. Thus, one might expect that the non-zero spectral points of the coset $(K_n) + \mathcal{G}$ of a compact sequence (K_n) are isolated. This expectation fail drastically as the following simple example shows.

EXAMPLE. Let (a_n) be an enumeration of the rational numbers in $[0, 1]$, and set
$$K_n := a_n P_n P_1 P_n = \text{diag}\,(a_n, 0, \ldots, 0).$$
The sequence (K_n) is compact (it consists of rank one matrices), but the spectrum of its coset $(K_n) + \mathcal{G}$ is the interval $[0, 1]$. $\square$

For later use we provide a technical result on strong limits of compact sequences. Its proof is based on the following lemma by Böttcher and Grudsky (Lemma 5.7 in [**11**]).

LEMMA 4.1. *Let H be a separable Hilbert space, and let (F_n) be a bounded sequence of operators $F_n \in L(H)$ with $r := \sup \text{rank}\,F_n < \infty$. Then there is an operator $F \in L(H)$ with $\text{rank}\,F \leq r$ such that, for each pair of vectors $x, y \in H$, the number $\langle Fx, y \rangle$ is a partial limit of the sequence $(\langle F_n x, y \rangle)_{n \in \mathbb{N}}$.*

PROPOSITION 4.2. *(a) Let $(K_n) \in \mathcal{K}$ be a strongly convergent sequence of essential rank r. Then its strong limit is an operator with rank at most r.*
(b) Let $(K_n) \in \mathcal{K}$ be a strongly convergent sequence. Then its strong limit is compact.

PROOF. We only prove assertion (b). Assertion (a) will follow from that proof.
Let $(K_n) \in \mathcal{K}$ and $K := \text{s-lim}\,K_n P_n$. Given $\varepsilon > 0$, choose a sequence $(F_n) \in \mathcal{K}$ with
$$\sup_n \|K_n - F_n\| < \varepsilon \quad \text{and} \quad r := \sup_n \text{rank}\,F_n < \infty.$$
By Lemma 4.1, there is an operator F with $\text{rank}\,F \leq r$ such that, for each pair of vectors $x, y \in H$, there is a strongly monotonically increasing sequence $\eta : \mathbb{N} \to \mathbb{N}$ with
$$\langle F_{\eta(n)} x, y \rangle \to \langle Fx, y \rangle.$$
Since strong convergence implies weak convergence, one has
$$\begin{aligned}
|\langle Kx, y\rangle - \langle Fx, y\rangle| &= |\lim_{n\to\infty}\langle K_{\eta(n)}x, y\rangle - \lim_{n\to\infty}\langle F_{\eta(n)}x, y\rangle| \\
&\leq \limsup_{n\to\infty} |\langle (K_{\eta(n)} - F_{\eta(n)})x, y\rangle| \\
&\leq \sup_{n\in\mathbb{N}} \|K_{\eta(n)} - F_{\eta(n)}\| < \varepsilon.
\end{aligned}$$

Consequently,
$$\|K - F\| = \sup_{\|x\|\leq 1, \|y\|\leq 1} |\langle Kx, y\rangle - \langle Fx, y\rangle| \leq \varepsilon.$$

Thus, each neighborhood of K contains an operator of finite rank, whence the compactness of K. □

We proceed with several equivalent characterizations of compact sequences.

4.2. Characterization via singular values

Recall that the singular values of an $n \times n$ matrix A are the non-negative square roots of the eigenvalues of A^*A. We denote them by

(4.2) $$\|A\| = \Sigma_1(A) \geq \Sigma_2(A) \geq \ldots \geq \Sigma_n(A) \geq 0$$

if they are ordered decreasingly and by

(4.3) $$0 \leq \sigma_1(A) \leq \sigma_2(A) \leq \ldots \leq \sigma_n(A) = \|A\|$$

in case of increasing order. Thus, $\sigma_k(A) = \Sigma_{n-k+1}(A)$. Notice that A^*A and AA^* are unitarily equivalent, whence $\Sigma_k(A) = \Sigma_k(A^*)$. We will also need the fact that every squared matrix A has a *singular value decomposition*

$$A = E^* \operatorname{diag}(\Sigma_1(A), \ldots, \Sigma_n(A))F$$

with unitary matrices E and F. To put these notions into a more general context, recall that the *kth approximation number* of a bounded linear operator A acting on a Hilbert space H is defined by

$$\Sigma_k(A) := \inf\{\|A - F\| : F \in L(H), \operatorname{rank} F < k\}.$$

If one identifies the $n \times n$-matrix A with the operator $P_n A P_n$ acting on $l^2(\mathbb{N})$, then one easily checks that the kth singular value $\Sigma_k(A)$ of A coincides with the kth approximation number of $P_n A P_n$ which justifies the notation. One can further show that, for each operator $A \in L(H)$,

(4.4) $$\Sigma_k(A) \to \|A\|_{ess} \quad \text{as} \quad k \to \infty$$

where the *essential norm* $\|A\|_{ess}$ of A is defined by $\inf\{\|A - K\| : K \in K(H)\}$, i.e., the essential norm of A is the norm of the coset $A + K(H)$ in the Calkin algebra $L(H)/K(H)$. Another basic fact established in [10] is that

(4.5) $$\lim_{n \to \infty} \Sigma_k(P_n A P_n) = \Sigma_k(A)$$

for each bounded linear operator A acting on a separable Hilbert space and for each positive integer k. If A is compact, this result is trivial.

We start with two simple technical lemmas which will be used several times.

LEMMA 4.3. *Let A be an $n \times n$ matrix with $\Sigma_r(A) > 0$ for some $r \in \{1, \ldots, n\}$. Then $\operatorname{rank}(A') \geq r$ for each $n \times n$ matrix A' with $\|A - A'\| < \Sigma_r(A)$.*

PROOF. Let E and F be unitary matrices such that

$$A = E^* \operatorname{diag}(\Sigma_1(A), \ldots, \Sigma_n(A))F$$

is the singular value decomposition of A. For $r \leq n$, the operator

$$P_r E A F^* P_r = \operatorname{diag}(\Sigma_1(A), \ldots, \Sigma_r(A), 0, \ldots, 0),$$

considered as acting on $\operatorname{im} P_r$, is invertible, and
$$\|(P_r EAF^* P_r|_{\operatorname{im} P_r})^{-1}\| = 1/\Sigma_r(A).$$
If A' is a matrix with $\|A - A'\| < \Sigma_r(A)$, then
$$\begin{aligned}\|P_r EAF^* P_r|_{\operatorname{im} P_r} - P_r EA'F^* P_r|_{\operatorname{im} P_r}\| &\leq \|A - A'\| < \Sigma_r(A) \\ &= \|(P_r EAF^* P_r|_{\operatorname{im} P_r})^{-1}\|^{-1}.\end{aligned}$$
By Neumann series, the operator $P_r EA'F^* P_r$ is invertible on the r-dimensional space $\operatorname{im} P_r$. Hence, $\operatorname{rank} A' \geq r$. □

LEMMA 4.4. *Let $(K_n) \in \mathcal{F}$ be a sequence with $\limsup_{n \to \infty} \Sigma_r(K_n) > 0$ for some $r \in \mathbb{N}$. Then* ess $\operatorname{rank}(K'_n) \geq r$ *for each sequence $(K'_n) \in \mathcal{F}$ with*
$$\|(K_n) - (K'_n)\| < \limsup_{n \to \infty} \Sigma_r(K_n).$$

PROOF. Let (K_n) and (K'_n) be as in the lemma and choose $C > 0$ such that
$$\|(K_n) - (K'_n)\| < C < \limsup_{n \to \infty} \Sigma_r(K_n).$$
By hypothesis, there is a subsequence (K_{n_k}) of (K_n) such that
$$\Sigma_r(K_{n_k}) > C \quad \text{for all } k \geq 1. \tag{4.6}$$
Contrary to what we want to show, we assume that ess $\operatorname{rank}(K'_n) < r$. Then there are sequences $(G_n) \in \mathcal{G}$ and $(K''_n) \in \mathcal{F}$ with $\operatorname{rank} K''_n < r$ for all n such that $(K'_n) = (K''_n) + (G_n)$. Since one still has $\|K_n - K''_n\| < C$ for all sufficiently large n, we conclude from (4.6) and from Lemma 4.3 that $\operatorname{rank} K''_{n_k} \geq r$ for all sufficiently large k. Contradiction. □

The announced characterization of compact sequences in terms of singular values reads as follows.

THEOREM 4.5. *The following conditions are equivalent for a sequence $(K_n) \in \mathcal{F}$:*
(a) $\lim_{k \to \infty} \sup_{n \geq k} \Sigma_k(K_n) = 0$;
(b) $\lim_{k \to \infty} \limsup_{n \to \infty} \Sigma_k(K_n) = 0$;
(c) *the sequence (K_n) is compact.*

Since the sequence $k \mapsto \sup_{n \geq k} \Sigma_k(K_n)$ is monotonically decreasing, the limit in (a) and (b) can be replaced by an infimum. A generalization of Theorem 4.5 will be derived in Theorem 5.2 below.

PROOF. The implication $(a) \Rightarrow (b)$ is evident. Let $(K_n) \in \mathcal{F}$ be a sequence which satisfies condition (b), and let (with unitary matrices E_n, F_n)
$$K_n = E_n^* \operatorname{diag}(\Sigma_1(K_n), \ldots, \Sigma_{\delta(n)}(K_n)) F_n$$
be the singular value decomposition of K_n. For every $n \in \mathbb{N}$ and $k \geq 1$, set
$$K_n^{(k)} := \begin{cases} E_n^* \operatorname{diag}(\Sigma_1(K_n) \ldots, \Sigma_{k-1}(K_n), 0, \ldots, 0) F_n & \text{if } 1 < k \leq n \\ 0 & \text{if } 1 = k \leq n \\ K_n & \text{if } n < k. \end{cases}$$
Then, for $n > k$,
$$\|K_n - K_n^{(k)}\| = \|E_n^* \operatorname{diag}(0, \ldots, 0, \Sigma_k(K_n), \ldots, \Sigma_{\delta(n)}(K_n)) F_n\| = \Sigma_k(K_n),$$

and the limsup formula (2.6) for the norm in $\mathcal{F}/\mathcal{G}$ yields
$$\|(K_n) - (K_n^{(k)}) + \mathcal{G}\|_{\mathcal{F}/\mathcal{G}} = \limsup_{n \to \infty} \Sigma_k(K_n).$$

Together with property (b), this implies that
$$\lim_{k \to \infty} \|(K_n) - (K_n^{(k)}) + \mathcal{G}\|_{\mathcal{F}/\mathcal{G}} = \lim_{k \to \infty} \limsup_{n \to \infty} \Sigma_k(K_n) = 0.$$

Thus, for each $k \in \mathbb{N}$, there is a sequence $(C_n^{(k)})$ in $\mathcal{G}$ such that
$$\lim_{k \to \infty} \|(K_n) - (K_n^{(k)}) - (C_n^{(k)})\|_{\mathcal{F}} = 0,$$

i.e., the sequence (K_n) is the limit as $k \to \infty$ of the sequences $(K_n^{(k)} + C_n^{(k)})_{n \in \mathbb{N}}$. Each of these sequences belongs to $\mathcal{K}$ since $\mathcal{G} \subset \mathcal{K}$ and since rank $K_n^{(k)} \leq k - 1$ by definition. Hence, (K_n) is a compact sequence.

For the implication (c) $\Rightarrow$ (a), let (K_n) be a compact sequence. Then the sequence $(\sup_{n \geq k} \Sigma_k(K_n))_{k \geq 1}$ is monotonically decreasing and bounded below by zero, hence, convergent. Assume that the limit of this sequence is positive. Then there is a $C > 0$ such that
$$\sup_{n \geq k} \Sigma_k(K_n) > C \quad \text{for all } k \geq 1.$$

Thus, for each $k \geq 1$, there is an $n_k \geq k$ such that

(4.7) $$\Sigma_k(K_{n_k}) \geq C \quad \text{for all } k \geq 1.$$

Further, since (K_n) is compact, there is a sequence $(R_n) \in \mathcal{F}$ with

(4.8) $$\sup_n \operatorname{rank} R_n < \infty \quad \text{and} \quad \sup_n \|K_n - R_n\| < C.$$

In particular, for each k one has
$$\|K_{n_k} - R_{n_k}\| < C,$$

which implies via Lemma 4.3 and (4.7) that rank $R_{n_k} \geq k$. Since k can be chosen arbitrarily large, this contradicts the first condition in (4.8). Hence, the sequence $(\sup_{n \geq k} \Sigma_k(K_n))_{k \geq 1}$ cannot have a positive limit, whence condition (a). $\square$

In the same vein one can prove the following characterization of sequences of essential rank r.

COROLLARY 4.6. *A sequence $(K_n) \in \mathcal{F}$ is of essential rank r if and only if*
$$\limsup_{n \to \infty} \Sigma_r(K_n) > 0 \quad \text{and} \quad \lim_{n \to \infty} \Sigma_{r+1}(K_n) = 0.$$

Together with Lemma 4.4 this corollary implies the *lower semi-continuity* of the essential rank function.

COROLLARY 4.7. *If* ess rank $(K_n) = r$, *then* ess rank $(K_n') \geq r$ *for all sequences (K_n') which are sufficiently close to (K_n).*

Another corollary concerns the behavior of the small singular values of K_n.

COROLLARY 4.8. *Let $(K_n) \in \mathcal{K}$. Then the limit $\lim_{n \to \infty} \sigma_k(K_n)$ exists and is equal to 0 for every k.*

PROOF. Let $\varepsilon > 0$. By Theorem 4.5, there is a k_0 such that $\sup_{n \geq k_0} \Sigma_{k_0}(K_n) < \varepsilon$. Then, for all $n \geq n_0 := k_0 + k - 1$,

$$\sigma_k(K_n) = \Sigma_{\delta(n)-k+1}(K_n) \leq \Sigma_{k_0}^{(n)} \leq \sup_{n \geq k_0} \Sigma_{k_0}(K_n) < \varepsilon,$$

which gives the assertion. □

4.3. Central rank characterizations

Let $\mathcal{C}$ be a C^*-algebra with identity e. Recall that the *center* $\operatorname{Cen} \mathcal{C}$ of $\mathcal{C}$ is the set of all elements of $\mathcal{C}$ which commute with each element of $\mathcal{C}$. Thus, $\operatorname{Cen} \mathcal{C}$ is a commutative C^*-subalgebra of $\mathcal{C}$. A nonzero element $k \in \mathcal{C}$ is said to be of *central rank one* if, for each $a \in \mathcal{C}$, there is a $c \in \operatorname{Cen} \mathcal{C}$ with $kak = ck$, and $k \in \mathcal{C}$ is said to be of *finite central rank* if it is the sum of a finite number of elements of central rank one. One easily checks that the product of an element of central rank one with an arbitrary element of $\mathcal{C}$ is either zero or of central rank one again. Thus, the elements of finite central rank form an ideal of $\mathcal{C}$. We refer to the elements of $\mathcal{C}$ which belong to the closure of that ideal as being *centrally compact*.

Let $k \in \mathcal{C}$ be of finite central rank. If $k = 0$, we define the central rank of k to be zero. If $k \neq 0$, there is a smallest positive integer r such that k can be written as the sum of r elements of central rank one. In this case we take r as the *central rank* of k and write

$$\operatorname{cen rank} k = r.$$

For the desired characterization of $\mathcal{K}$ as centrally compact sequences we describe the centers of the algebras $\mathcal{F}$ and $\mathcal{F}/\mathcal{G}$.

THEOREM 4.9. *(a) The center of $\mathcal{F}$ consists of all sequences $(\alpha_n I_n)$ with $(\alpha_n) \in l^\infty(\mathbb{N})$.*
(b) The center of $\mathcal{F}/\mathcal{G}$ consists of all cosets modulo $\mathcal{G}$ of sequences $(\alpha_n I_n)$ with $(\alpha_n) \in l^\infty(\mathbb{N})$.

Assertion (a) implies a natural *-isomorphism between the center of $\mathcal{F}$ and the C^*-algebra $l^\infty(\mathbb{N})$ and, since $l^\infty(\mathbb{N}) \cap \mathcal{G} = c_0(\mathbb{N})$, assertion (b) establishes a *-isomorphism between the center of $\mathcal{F}/\mathcal{G}$ and the quotient algebra $l^\infty(\mathbb{N})/c_0(\mathbb{N})$.

PROOF. The proof of assertion (a) is evident since the center of $\mathbb{C}^{n \times n}$ consists of the multiples of the identity matrix only. It is also evident that every coset $(\alpha_n I_n) + \mathcal{G}$ with $(\alpha_n) \in l^\infty(\mathbb{N})$ belongs to the center of $\mathcal{F}/\mathcal{G}$.

Let, conversely, $(A_n) + \mathcal{G}$ be a coset in the center of $\mathcal{F}/\mathcal{G}$. We have to show that $(A_n) + \mathcal{G}$ is of the form $(\alpha_n I_n) + \mathcal{G}$ with $(\alpha_n) \in l^\infty(\mathbb{N})$. Since every element of the commutative C^*-algebra $\operatorname{Cen}(\mathcal{F}/\mathcal{G})$ can be written as a linear combination of four positive elements, it remains to prove this implication in case $(A_n) + \mathcal{G}$ is a positive element in the center of $\mathcal{F}/\mathcal{G}$. Since closed ideals in C^*-algebras lift positive elements (see, e.g., [52], 2.2.10), we can further assume that (A_n) is a positive element of $\mathcal{F}$, i.e., that all matrices A_n are positive.

Write each $A_n \geq 0$ as $E_n^* D_n E_n$ with a unitary matrix E_n and a diagonal matrix

$$D_n = \operatorname{diag}(\lambda_1(A_n), \lambda_2(A_n), \ldots, \lambda_{\delta(n)}(A_n))$$

with the eigenvalues $\lambda_1(A_n) \leq \lambda_2(A_n) \leq \ldots \leq \lambda_{\delta(n)}(A_n) = \|A_n\|$ of A_n. Further, define a sequence $(G_n) \in \mathcal{F}$ by $A_n = \lambda_1(A_n) I_n + G_n$. Since the sequence $(\lambda_1(A_n) I_n)$

belongs to the center of $\mathcal{F}$, the coset $(G_n) + \mathcal{G}$ lies in the center of $\mathcal{F}/\mathcal{G}$. Since the mapping
$$(B_n) + \mathcal{G} \mapsto (E_n B_n E_n^*) + \mathcal{G}$$
defines an automorphism of $\mathcal{F}/\mathcal{G}$, the sequence D_n' with
$$D_n' := E_n G_n E_n^* = \operatorname{diag}(0, \lambda_2(A_n) - \lambda_1(A_n), \ldots, \lambda_{\delta(n)}(A_n) - \lambda_1(A_n))$$
belongs to the center of $\mathcal{F}/\mathcal{G}$, too. Consider the matrix $C_n := (c_{ij})_{i,j=1}^{\delta(n)}$ with $c_{1n} = 1$ and $c_{ij} = 0$ for all other entries. The sequence (C_n) is evidently bounded; so it belongs to $\mathcal{F}$, and it commutes with (D_n') modulo a zero sequence,
$$\|C_n D_n' - D_n' C_n\| \to 0 \quad \text{as } n \to \infty.$$
From
$$C_n D_n' - D_n' C_n = (\lambda_{\delta(n)}(A_n) - \lambda_1(A_n)) C_n$$
and $\|C_n\| = 1$ one concludes that $\lambda_{\delta(n)}(A_n) - \lambda_1(A_n) \to 0$. Since
$$\lambda_{\delta(n)}(A_n) - \lambda_1(A_n) \geq \max_{1 \leq i \leq \delta(n)} (\lambda_i(A_n) - \lambda_1(A_n)) = \|D_n'\|,$$
this implies $\|D_n'\| \to 0$ and, hence, $(G_n) \in \mathcal{G}$. $\square$

THEOREM 4.10. *Let $(K_n) \in \mathcal{F}$.*
(a) *The sequence (K_n) is of central rank one in $\mathcal{F}$ if and only if $(K_n) \neq 0$ and if rank $K_n \leq 1$ for each n.*
(b) *The coset $(A_n) + \mathcal{G}$ is of central rank one in $\mathcal{F}/\mathcal{G}$ if and only if $(A_n) \notin \mathcal{G}$ and if there is a sequence $(K_n) \in \mathcal{F}$ of matrices K_n of rank less than or equal to one such that $(A_n) - (K_n) \in \mathcal{G}$.*

Notice that, in the setting of assertion (b), infinitely many of the matrices K_n must have rank one.

PROOF. (a) If (K_n) is of central rank one in $\mathcal{F}$, then every matrix K_n is of rank one in $\mathbb{C}^{\delta(n) \times \delta(n)}$. Hence, rank $K_n \leq 1$ for all n.

Conversely, let $(K_n) \in \mathcal{F}$ be a sequence of matrices of rank one. Then, for every sequence $(A_n) \in \mathcal{F}$, there is a bounded sequence (α_n) of complex numbers such that
$$K_n A_n K_n = \alpha_n K_n \quad \text{for all } n.$$
For the boundedness of (α_n) observe that one can choose $\alpha_n = 0$ if $K_n = 0$ and that $|\alpha_n| \leq \|K_n\| \|A_n\|$ in case $K_n \neq 0$. Thus, (K_n) is of central rank one.
(b) Let $(A_n) + \mathcal{G}$ be of central rank one in $\mathcal{F}/\mathcal{G}$. Consider the singular value decomposition
$$A_n = E_n^* \Sigma_n F_n$$
of A_n with unitary matrices E_n and F_n and with the diagonal matrix
$$\Sigma_n := \operatorname{diag}(\sigma_1(A_n), \sigma_2(A_n), \ldots, \sigma_{\delta(n)}(A_n))$$
containing the increasingly ordered singular values of A_n. Set
$$B_n := F_n^* \operatorname{diag}(0, 0, \ldots, 0, 1) E_n.$$
By Theorem 4.9 and by the definition of a central rank one element, there is a sequence $(\beta_n) \in l^\infty(\mathbb{N})$ such that
$$(A_n)(B_n)(A_n) - (\beta_n)(A_n) \in \mathcal{G}$$

or, equivalently,
$$\|\Sigma_n \mathrm{diag}\,(0, 0, \ldots, 0, 1)\Sigma_n - \beta_n \Sigma_n\| \to 0.$$

For the last two entries on the main diagonal, this implies that

(4.9)
$$\sigma_{\delta(n)}(A_n)^2 - \beta_n \sigma_{\delta(n)}(A_n) \to 0$$

and

(4.10)
$$\beta_n \sigma_{\delta(n)-1}(A_n) \to 0.$$

Assume for a moment that $\limsup \sigma_{\delta(n)-1}(A_n) > 0$. Then there are a constant $C > 0$ and a strongly monotonically increasing sequence $\eta : \mathbb{N} \to \mathbb{N}$ such that

(4.11)
$$\sigma_{\delta(\eta(n))-1}(A_{\eta(n)}) \geq C \quad \text{for each } n \in \mathbb{N}.$$

By (4.10), one has $\beta_{\eta(n)} \to 0$, which implies via (4.9) that $\sigma_{\delta(\eta(n))}(A_{\eta(n)}) \to 0$. Since $\sigma_{\delta(\eta(n))-1}(A_{\eta(n)}) \leq \sigma_{\delta(\eta(n))}(A_{\eta(n)})$ we obtain $\sigma_{\delta(\eta(n))-1}(A_{\eta(n)}) \to 0$ which contradicts (4.11). Hence,

(4.12)
$$\lim_{n\to\infty} \sigma_{\delta(n)-1}(A_n) = 0.$$

Now write Σ_n as $\Sigma'_n + \Sigma''_n$ with
$$\Sigma'_n := \mathrm{diag}\,(\sigma_1(A_n), \sigma_2(A_n), \ldots, \sigma_{\delta(n)-1}(A_n), 0).$$

By (4.12), the sequence (Σ'_n) tends to zero in the norm. Thus,
$$A_n = E_n^* \Sigma_n F_n = E_n^* \Sigma'_n F_n + E_n^* \Sigma''_n F_n$$

yields a decomposition of (A_n) into sum of a sequence in $\mathcal{G}$ and a sequence of rank one matrices. Finally, the sequence $K_n := E_n^* \Sigma''_n F_n$ cannot belong to $\mathcal{G}$ if $(A_n) \notin \mathcal{G}$.

The reverse assertion is easily checked. Let (K_n) be a sequence of rank one matrices which is not in $\mathcal{G}$. Then, for each sequence $(A_n) \in \mathcal{F}$, there is a sequence (α_n) of complex numbers such that
$$K_n A_n K_n = \alpha_n K_n \quad \text{for each } n \in \mathbb{N}.$$

If $K_n = 0$, we choose $\alpha_n = 0$. Otherwise, α_n is uniquely determined, and the estimate
$$|\alpha_n|\,\|K_n\| = \|\alpha_n K_n\| = \|K_n A_n K_n\| \leq \|K_n\|^2 \|A_n\|$$

gives
$$|\alpha_n| \leq \|(A_n)\|_{\mathcal{F}} \|(K_n)\|_{\mathcal{F}}.$$

Thus, the sequence (α_n) is bounded, which implies that $(K_n) + \mathcal{G}$ is of central rank one in $\mathcal{F}/\mathcal{G}$. □

COROLLARY 4.11. *The ideal $\mathcal{K}$ coincides with the ideal of the centrally compact elements of the algebra $\mathcal{F}$, and $\mathcal{K}/\mathcal{G}$ is the ideal of the centrally compact elements of $\mathcal{F}/\mathcal{G}$.*

THEOREM 4.12. *A sequence $\mathbf{A} \in \mathcal{F}$ is of finite essential rank if and only if the coset $\mathbf{A} + \mathcal{G}$ is of finite central rank in $\mathcal{F}/\mathcal{G}$. In this case,*
$$\mathrm{ess\,rank}\,\mathbf{A} = \mathrm{cen\,rank}\,(\mathbf{A} + \mathcal{G}).$$

PROOF. Let $\mathbf{A}+\mathcal{G}$ be of finite central rank r in $\mathcal{F}/\mathcal{G}$. Then there are sequences $(K_n^{(1)})$, ..., $(K_n^{(r)})$ if $\mathcal{F}$ such that each coset $(K_n^{(l)}) + \mathcal{G}$ is of central rank 1 and

$$\mathbf{A} - (K_n^{(1)}) - \ldots - (K_n^{(r)}) \in \mathcal{G}.$$

By Theorem 4.10, these sequences can be chosen such that

$$\operatorname{rank} K_n^{(l)} \leq 1 \quad \text{for all } l = 1, \ldots, r \text{ and } n \in \mathbb{N}.$$

Thus, $\mathbf{A} = (K_n^{(1)} + \ldots + K_n^{(r)}) + (G_n)$ with a sequence $(G_n) \in \mathcal{G}$ and with

$$\operatorname{rank}(K_n^{(1)} + \ldots + K_n^{(r)}) \leq r.$$

Hence, $\mathbf{A}$ is of finite essential rank, and

(4.13) $$\operatorname{ess rank} \mathbf{A} \leq r = \operatorname{cen rank}(\mathbf{A} + \mathcal{G}).$$

Conversely, let $\mathbf{A} \in \mathcal{F}$ be a sequence of finite essential rank. Then we find sequences $(K_n) \in \mathcal{F}$ and $(G_n) \in \mathcal{G}$ with

$$\mathbf{A} = (K_n) + (G_n) \quad \text{and} \quad \operatorname{rank} K_n \leq r.$$

For each $n \in \mathbb{N}$, we choose an orthonormal basis $e_1^{(n)}, \ldots, e_{r_n}^{(n)}$ in $\operatorname{im} K_n$ and, for $l = 1, \ldots, r_n$, we write $P_l^{(n)}$ for the orthogonal projection of $\operatorname{im} K_n$ onto $\mathbb{C}e_l^{(n)}$. In case $r_n < r$, then we set $P_l^{(n)} = 0$ for $l = r_n + 1, \ldots, r$. Then, evidently, $P_1^{(n)} + \ldots + P_r^{(n)} = I$ on $\operatorname{im} K_n$ and $\operatorname{rank} P_l^{(n)} \leq 1$ for all $l \leq r$. Thus, the sequence (K_n) can be written as the sum of the sequences

$$(P_l^{(n)} K_n)_{n \in \mathbb{N}} \quad \text{with } l = 1, \ldots, r.$$

Each of these sequences belongs to $\mathcal{F}$ and is a sequence of rank one matrices. Hence, each coset $(P_l^{(n)} K_n) + \mathcal{G}$ is an element of $\mathcal{F}/\mathcal{G}$ of central rank at most one. This implies that the coset $\mathbf{A} + \mathcal{G} = (K_n) + \mathcal{G}$ is of finite central rank in $\mathcal{F}/\mathcal{G}$ and that

$$\operatorname{cen rank}(\mathbf{A} + \mathcal{G}) \leq r = \operatorname{ess rank} \mathbf{A}.$$

Together with (4.13), this gives the assertion. $\square$

Notice that the characterization of sequences in $\mathcal{K}$ as centrally compact in the algebra $\mathcal{F}$ implies another notion of central rank. We will see later on that it is the central rank of the coset $(A_n) + \mathcal{G}$ rather than the central rank of the sequence (A_n) itself which appears in many applications.

4.4. Minimal and maximal characterizations

When thinking about reasonable ways to define an appropriate ideal of compact sequences, one certainly wants to have the constant sequence (P_1) to be compact. So one comes to consider the smallest closed ideal of $\mathcal{F}$ which contains this sequence, and this ideal is a minimal candidate of what might be called an ideal of compact sequences. At the other end of the scale, one might call a sequence $\mathbf{K} = (K_n) \in \mathcal{F}$ compact if $W(\mathbf{K})$ is a compact operator for each irreducible representation of $\mathcal{F}$, yielding a maximal candidate for an ideal of compact sequences. The following theorems show that both the minimal and the maximal candidate already coincide with the ideal $\mathcal{K}$.

THEOREM 4.13. *$\mathcal{K}$ is the smallest closed ideal of $\mathcal{F}$ which contains the constant sequence (P_1).*

4.4. MINIMAL AND MAXIMAL CHARACTERIZATIONS

PROOF. Let, for a moment, $\mathcal{K}'$ stand for the smallest closed ideal of $\mathcal{F}$ which contains the sequence (P_1). Evidently, $\mathcal{K}' \subseteq \mathcal{K}$. For the reverse inclusion, we have to verify that every sequence of rank one matrices belongs to $\mathcal{K}'$.

Let (K_n) be a bounded sequence of matrices with rank $K_n \leq 1$, and define a sequence (K'_n) by

$$K'_n := \begin{cases} \|K_n^* K_n\|^{-1} K_n^* K_n & \text{if } K_n \neq 0 \\ P_1 & \text{if } K_n = 0. \end{cases}$$

We claim that every matrix K'_n is an orthogonal projection of rank 1 and that $K_n K'_n = K_n$. There is nothing to prove if $K_n = 0$. Let rank $K_n = 1$. Then there is a uniquely determined complex number α_n such that

(4.14) $$K_n K_n^* K_n = \alpha_n K_n$$

and, thus,

(4.15) $$K_n^* K_n K_n^* K_n = \alpha_n K_n^* K_n.$$

Since $K_n^* K_n$ is non-zero and positive, one has $\alpha_n > 0$. Thus, taking norms in (4.15) yields

$$\|K_n^* K_n\|^2 = \alpha_n \|K_n^* K_n\|,$$

whence $\alpha_n = \|K_n^* K_n\|$. Dividing (4.15) by $\|K_n^* K_n\|$ shows that K'_n is an orthogonal projection of rank one, and dividing (4.14) by $\|K_n^* K_n\|$ shows that $K_n K'_n = K_n$. This proves the claim. Consequently, there are unitary matrices E_n such that

$$K'_n = E_n^* \operatorname{diag}(1, 0, \ldots, 0) E_n = E_n^* P_1 E_n.$$

The sequence (E_n) is bounded and belongs therefore to $\mathcal{F}$. Hence, the sequence $(K'_n) = (E_n^*)(P_1)(E_n)$ belongs to $\mathcal{K}'$. Because of $(K_n) = (K_n)(K'_n)$, this implies that $(K_n) \in \mathcal{K}'$, too. □

THEOREM 4.14. (a1) Let $\mathbf{K} \in \mathcal{F}$ be a sequence of rank one matrices and W an irreducible representation of $\mathcal{F}$. Then $\operatorname{rank} W(\mathbf{K}) \leq 1$.

(a2) Let $\mathbf{K} + \mathcal{G} \in \mathcal{F}/\mathcal{G}$ be a coset of central rank one and W an irreducible representation of $\mathcal{F}/\mathcal{G}$. Then $\operatorname{rank} W(\mathbf{K} + \mathcal{G}) \leq 1$.

(b1) A sequence $\mathbf{K} \in \mathcal{F}$ belongs to the ideal $\mathcal{K}$ if and only if $W(\mathbf{K})$ is compact for every irreducible representation W of $\mathcal{F}$.

(b2) A coset $\mathbf{K} + \mathcal{G} \in \mathcal{F}/\mathcal{G}$ belongs to the ideal $\mathcal{K}/\mathcal{G}$ if and only if $W(\mathbf{K} + \mathcal{G})$ is compact for every irreducible representation W of $\mathcal{F}/\mathcal{G}$.

For the proof, we need two auxiliary results.

THEOREM 4.15. Let $\mathcal{C}$ be a C^*-algebra with identity, and let k be a central rank one element of $\mathcal{C}$. Then $W(k)$ is an operator with rank less than or equal to one for every irreducible representation W of $\mathcal{C}$.

PROOF. Let $W : \mathcal{C} \to L(H)$ be an irreducible representation of $\mathcal{C}$, and let k be a central rank one element of $\mathcal{A}$. If $W(k) = 0$, then nothing is to prove. So let $W(k) \neq 0$.

Let $x, \tilde{x}$ be vectors in $\operatorname{im} W(k)$ with $x \neq 0$, and choose vectors $y, \tilde{y} \in H$ such that $x = W(k)y$ and $\tilde{x} = W(k)\tilde{y}$. Since irreducible representations are algebraically cyclic, we have $\pi(\mathcal{A})x = H$. In particular, there is an element $a \in \mathcal{C}$ such that $W(a)W(k)y = W(a)x = \tilde{y}$ and, hence, $W(k)W(a)W(k)y = W(k)\tilde{y}$. Moreover, since k is of central rank one, there is an element c in the center of $\mathcal{C}$ such that $kak =$

ck. Since irreducible representations map central elements to scalar operators, we have
$$W(k)W(a)W(k) = \mu W(k) \quad \text{with} \quad \mu = W(c) \in \mathbb{C}.$$
Hence, $\mu W(k)y = W(k)\tilde{y}$, i.e., $\mu x = \tilde{x}$. This shows that $\operatorname{im} W(k) = \mathbb{C}x$. In particular, $W(k)$ has rank one. □

The following is taken from [27], Proposition 4.1.8.

THEOREM 4.16. *Let $\mathcal{A}$ be a C^*-subalgebra of a C^*-algebra $\mathcal{B}$ and $W : \mathcal{A} \to L(K)$ an irreducible representation of $\mathcal{A}$. Then there exist an irreducible representation $\pi : \mathcal{B} \to L(H)$, a closed subspace H_1 of the Hilbert space H, and a bijective isometry $U : H_1 \to K$ such that*
$$W(a) = U\pi(a)|_{H_1} U^* \quad \text{for all } a \in \mathcal{A}.$$

PROOF OF THEOREM 4.14. By Theorem 4.10, assertions (a1) and (a2) become special cases of Theorem 4.15. Further, since the irreducible representations of $\mathcal{F}/\mathcal{G}$ are just the irreducible representations of $\mathcal{F}$ which have $\mathcal{G}$ in their kernel, assertion (b2) follows from (b1). So we are left with verifying assertion (b1).

Let $\mathcal{K}'$ stand for the set of all sequences $\mathbf{K}$ in $\mathcal{F}$ such that $W(\mathbf{K})$ is compact for every irreducible representation W of $\mathcal{F}$. The inclusion $\mathcal{K} \subseteq \mathcal{K}'$ is an immediate consequence of assertion (a1). For the proof of the reverse inclusion, we will show that, for every sequence $\mathbf{K} \in \mathcal{F} \setminus \mathcal{K}$, there is an irreducible representation W of $\mathcal{F}$ such that $W(\mathbf{K})$ is not compact.

Let $\mathbf{K} = (K_n) \in \mathcal{F}$ be a sequence which is not in $\mathcal{K}$, and let $\operatorname{Id} \mathbf{K}$ refer to the smallest closed ideal of $\mathcal{F}$ which contains the sequence $\mathbf{K}$. Further, let
$$E_n^* K_n^* K_n E_n = \operatorname{diag}(\Sigma_1(K_n)^2, \Sigma_2(K_n)^2, \ldots, \Sigma_{\delta(n)}(K_n)^2)$$
with a unitary matrix E_n and with decreasingly ordered singular values. Since $(K_n) \notin \mathcal{K}$, Theorem 4.5 implies that $\lim_{k \to \infty} \sup_{n \geq k} \Sigma_k(K_n)^2 \neq 0$. Thus, as we checked in the proof of Theorem 4.5, there is a $C > 0$ as well as a sequence $(n_k)_{k \geq 1}$ with $n_k \geq k$ such that

(4.16) $$\Sigma_k(K_{n_k})^2 \geq C \quad \text{for all } k.$$

The sequence (n_k) can be chosen strictly monotonically increasing. Indeed, suppose that we have already found $n_1 < n_2 < \ldots < n_k$. Choose $k' > n_k$ and a corresponding $n' \geq k'$ such that $\Sigma_{k'}(K_{n'})^2 \geq C$ which is possible due to (4.16). Then $n_{k+1} := n'$ satisfies $n_{k+1} = n' \geq n_k + 1 \geq k+1$, yielding the desired strict monotonicity.

So let (n_k) be a strictly monotonically increasing sequence satisfying (4.16). Set
$$L_n := \operatorname{diag}\left(\Sigma_1(K_{n_k})^{-2}, \ldots, \Sigma_k(K_{n_k})^{-2}, 0, \ldots, 0\right)$$
if $n = n_k$ for some k and $L_n := 0$ if $n \neq n_k$. The sequence (L_n) is bounded because of
$$\Sigma_j(K_{n_k})^{-2} \leq 1/C \quad \text{for } j = 1, \ldots, k.$$
Hence, the sequence $\mathbf{D} = (D_n)$ with
$$D_n := L_n E_n^* K_n^* K_n E_n = \begin{cases} 0 & \text{if } n \neq n_k \\ \operatorname{diag}(1, \ldots, 1, 0, \ldots, 0) & \text{if } n = n_k \end{cases}$$

with k ones and $\delta(n_k) - k$ zeros on the diagonal belongs to the ideal $\mathrm{Id}\,\mathbf{K}$. Then all sequences

(4.17) $$(D_n P_n A P_n D_n) \quad \text{with} \quad A \in L(l^2(\mathbb{N}))$$

lie in $\mathrm{Id}\,\mathbf{K}$, and we consider the smallest closed C^*-subalgebra $\mathcal{B}$ of $\mathrm{Id}\,\mathbf{K}$ which contains all sequences (4.17). Clearly, the strong limit s-$\lim B_{n_k}$ as $k \to \infty$ exists for every sequence $\mathbf{B} = (B_n) \in \mathcal{B}$ and, in particular,

$$\text{s-}\lim_{k\to\infty} D_{n_k} P_{n_k} A P_{n_k} D_{n_k} = A$$

for all sequences of the form (4.17). The mapping

$$W : \mathcal{B} \to L(l^2(\mathbb{N})), \quad \mathbf{B} \mapsto \text{s-}\lim_{k\to\infty} B_{n_k}$$

is an irreducible representation of $\mathcal{B}$ (all compact operators lie in the range of W) which maps (D_n) to the identity operator. From Theorem 4.16 we know that there exists an irreducible representation $\pi : \mathcal{F} \to L(H)$, a closed subspace H_1 of H, and a bijective isometry $U : H_1 \to l^2(\mathbb{N})$ such that

(4.18) $$W(\mathbf{B}) = U\pi(\mathbf{B})|_{H_1} U^* \quad \text{for all } \mathbf{B} \in \mathcal{B}.$$

The operator $\pi(\mathbf{K})$ cannot be compact. Indeed, suppose for a moment that $\pi(\mathbf{K})$ is a compact operator. Then the operators $\pi(\mathbf{R})$ are compact for all sequences $\mathbf{R} \in \mathrm{Id}\,\mathbf{K}$ whence the compactness of $\pi(\mathbf{D})$. Then, by (4.18), $W(\mathbf{D})$ must be compact. This is impossible since $W(\mathbf{D})$ is the identity operator on $l^2(\mathbb{N})$ as we already observed. Thus, $\mathbf{K} \notin \mathcal{K}'$, yielding the inclusion $\mathcal{K}' \subseteq \mathcal{K}$. $\square$

4.5. Compact sequences in $\mathcal{S}(\mathbb{N})$ and $\mathcal{S}(\mathbb{Z})$

Next we describe the compact sequences in the algebras $\mathcal{S}(\mathbb{N})$ and $\mathcal{S}(\mathbb{Z})$ of the finite sections method.

THEOREM 4.17. *The intersection $\mathcal{S}(\mathbb{N}) \cap \mathcal{K}$ consists of all sequences $(A_n) \in \mathcal{S}(\mathbb{N})$ which converge strongly to a compact operator. Equivalently, $\mathcal{S}(\mathbb{N}) \cap \mathcal{K}$ is the set of all sequences of the form $(P_n K P_n) + (K_n)$ with a compact operator K and a sequence $(K_n) \in \mathcal{K}_0(\mathbb{N})$.*

PROOF. If a sequence $(K_n) \in \mathcal{K}$ lies in $\mathcal{S}(\mathbb{N})$, then it converges strongly and its strong limit is compact by Proposition 4.2 (b). Conversely, if $(A_n) \in \mathcal{S}(\mathbb{N})$ is a sequence with compact strong limit K then, by Theorem 2.14,

$$(A_n) = (P_n K P_n) + (K_n) \quad \text{with} \quad (K_n) \in \mathcal{K}_0(\mathbb{N}).$$

Evidently, the sequence $(P_n K P_n)$ is compact, and it remains to show the compactness of the sequences in $\mathcal{K}_0(\mathbb{N})$. The latter set coincides with the quasicommutator ideal (Theorem 2.15) which is generated by the sequences

$$(P_n A B P_n) - (P_n A P_n)(P_n B P_n) = (P_n A Q_n B P_n)$$

with band operators A and B. It remains to verify the compactness of the sequence $(P_n A Q_n B P_n)$. Let $r \geq 0$ be the band width of B, i.e., the entries b_{ij} in the matrix representation of B with respect to the standard basis vanish for $|i-j| > r$. Looking at the matrix representation of $Q_n B P_n$, one easily realizes that this operator has rank at most r for each $n \in \mathbb{N}$. Thus, the sequence $(P_n A Q_n B P_n)$ is of essential rank not greater than r, whence its compactness. $\square$

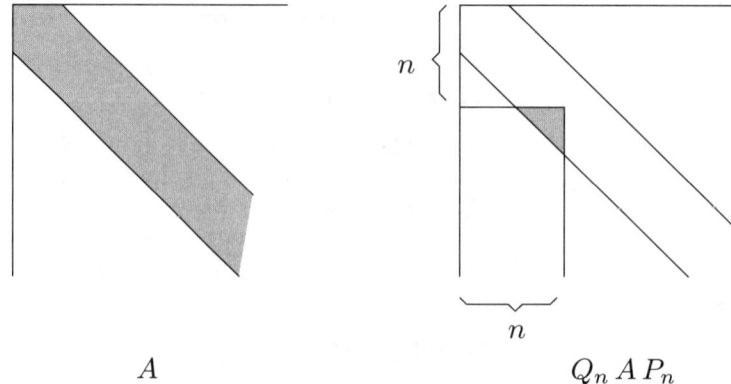

A $\qquad\qquad\qquad\qquad Q_n A P_n$

The following result determines the essential rank of an arbitrary compact sequence in $\mathcal{S}(\mathbb{N})$ in terms of the ranks of its limit operators. Recall that for a sequence $\mathbf{A} := (A_n) \in \mathcal{S}(\mathbb{N})$, the operator spectrum $\sigma_{op,1}(\mathbf{A})$ consists of all *-strong limits of subsequences of the sequence $(U_{-n} A_n U_n)_{n \geq 1}$. In case $\mathbf{A}$ is a compact sequence in $\mathcal{S}(\mathbb{N})$, then all operators in $\sigma_{op,1}(\mathbf{A})$ are compact due to Lemma 4.1.

THEOREM 4.18. *Let $K \in l^2(\mathbb{N})$ be a compact operator and let $\mathbf{K} := (K_n) \in \mathcal{K}_0$. Then the sequence $(P_n K P_n + K_n)$ is compact, and*

(4.19) $\quad \mathrm{ess\,rank}\,(P_n K P_n + K_n) = \mathrm{rank}\, K + \max\,\{\mathrm{rank}\,\mathbf{K}_h : \mathbf{K}_h \in \sigma_{op,1}(\mathbf{K})\}.$

We split the proof into several lemmas.

LEMMA 4.19. *Let $K \in L(l^2(\mathbb{N}))$ be compact and $(K_n) \in \mathcal{K}_0(\mathbb{N})$. Then*

(4.20) $\quad \mathrm{ess\,rank}\,(P_n K P_n + K_n) = \mathrm{ess\,rank}\,(P_n K P_n) + \mathrm{ess\,rank}\,(K_n).$

PROOF. It follows immediately from the definition of the essential rank that

$$\mathrm{ess\,rank}\,(A_n + B_n) \leq \mathrm{ess\,rank}\,(A_n) + \mathrm{ess\,rank}\,(B_n)$$

for arbitrary compact sequences (A_n) and (B_n). Thus we are left with verifying the estimate

$$\mathrm{ess\,rank}\,(P_n K P_n + K_n) \geq \mathrm{ess\,rank}\,(P_n K P_n) + \mathrm{ess\,rank}\,(K_n).$$

For $n \in \mathbb{N}$, let $[\sqrt{n}]$ refer to the largest integer which is not greater than $\sqrt{n}$. Since $P_{[\sqrt{n}]} \to I$ *-strongly and K is compact,

$$\|P_n K P_n - P_{[\sqrt{n}]} K P_{[\sqrt{n}]}\| \to 0 \qquad \text{as} \quad n \to \infty,$$

and since (K_n) is localized at the right-hand end of the interval $\{0, 1, \ldots, n-1\}$ due to Theorem 2.15 (d), one also has

$$\|K_n - (P_n - P_{n-[\sqrt{n}]}) K_n (P_n - P_{n-[\sqrt{n}]})\| \to 0 \qquad \text{as} \quad n \to \infty.$$

Since the essential rank is invariant under perturbations from the ideal $\mathcal{G}$ of the zero sequences, we conclude further that

$$\mathrm{ess\,rank}\,(P_n K P_n + K_n)$$
$$= \mathrm{ess\,rank}\,(P_{[\sqrt{n}]} K P_{[\sqrt{n}]} + (P_n - P_{n-[\sqrt{n}]}) K_n (P_n - P_{n-[\sqrt{n}]})).$$

Abbreviate $P_{[\sqrt{n}]}KP_{[\sqrt{n}]}$ to A_n and $(P_n - P_{n-[\sqrt{n}]})K_n(P_n - P_{n-[\sqrt{n}]})$ to B_n, and let ess rank $(A_n) =: k$ and ess rank $(B_n) =: l$. By Corollary 4.6, there is a $C > 0$ such that

$$\limsup_{n \to \infty} \Sigma_k(A_n) > C, \qquad \lim_{n \to \infty} \Sigma_{k+1}(A_n) = 0$$

and

$$\limsup_{n \to \infty} \Sigma_l(B_n) > C, \qquad \lim_{n \to \infty} \Sigma_{l+1}(B_n) = 0.$$

Choose $n_0 \in \mathbb{N}$ such that

$$\Sigma_{k+1}(A_n) < C/2, \quad \Sigma_{l+1}(B_n) < C/2 \quad \text{and} \quad \Sigma_k(A_n) > C$$

for all $n \geq n_0$ (the latter being possible by (4.5) and due to the special structure of the matrices A_n which are finite sections of the compact operator K), and choose a monotonically increasing sequence $(n_r)_{r \geq 1}$ with $n_1 \geq n_0$ such that

$$\Sigma_k(B_{n_r}) > C \quad \text{for each } r \geq 1.$$

Since $A_n + B_n$ has a diagonal block structure with A_n being located at the left upper and B_n at the right lower corner, the set of the non-zero singular values of $A_n + B_n$ is precisely the union of the sets of the non-zero singular values of the matrices A_n and B_n, each of these sets containing the singular values in accordance with their multiplicity. Thus, and due to our arrangements,

$$\Sigma_{k+l}(A_{n_r} + B_{n_r}) = \min\{\Sigma_k(A_{n_r}), \Sigma_l(B_{n_r})\} > C$$

for each $r \geq 1$. Consequently, $\limsup_{n \to \infty} \Sigma_{k+l}(A_n + B_n) \geq C > 0$, whence

$$\begin{aligned} \text{ess rank}\,(A_n + B_n) &\geq k + l = \text{ess rank}\,(A_n) + \text{ess rank}\,(B_n) \\ &= \text{ess rank}\,(P_n K P_n) + \text{ess rank}\,(K_n) \end{aligned}$$

by Corollary 4.6 and by the invariance of the essential rank under perturbations by zero sequences. $\square$

LEMMA 4.20. *If $K \in L(l^2(\mathbb{N}))$ is compact, then $(P_n K P_n)$ is a compact sequence, and*

$$\text{ess rank}\,(P_n K P_n) = \text{rank}\,K.$$

Indeed, for each $n \in \mathbb{N}$, one has rank $P_n K P_n \leq$ rank K, whence the estimate ess rank $(P_n K P_n) \leq$ rank K. The reverse estimate follows from Proposition 4.2 (*a*). $\square$

The following lemma completes the proof of Theorem 4.18.

LEMMA 4.21. *Let $\mathbf{K} := (K_n) \in \mathcal{K}_0$. Then*

(4.21) $$\text{ess rank}\,\mathbf{K} = \max\{\text{rank}\,\mathbf{K}_h : \mathbf{K}_h \in \sigma_{op,1}(\mathbf{K})\}.$$

PROOF. Let ess rank $\mathbf{K} =: r$. It follows from Lemma 4.1 and Proposition 4.2 that rank $\mathbf{K}_h \leq r$ for each limit operator $\mathbf{K}_h \in \sigma_{op,1}(\mathbf{K})$. Thus, we have to establish the existence of a limit operator of $\mathbf{K}$ for which rank $\mathbf{K}_h \geq r$.

By Theorem 2.15 (*d*) and by the lower semi-continuity of the essential rank (Corollary 4.7), there is an n_0 such that the sequence $\mathbf{K}^{(1)} = (K_n^{(1)})$ with

$$K_n^{(1)} := (P_n - P_{n-n_0})K_n(P_n - P_{n-n_0})$$

has essential rank $l \geq r$. In fact $\mathbf{K}^{(1)}$ has essential rank equal to r since it is a product of the essential rank r sequence $\mathbf{K}$ by other sequences. Moreover, we can choose $n_0 > r$.

Since $\mathbf{K}^{(1)}$ has essential rank r, there is a monotonically increasing sequence $h : \mathbb{N} \to \mathbb{N}$ such that

(4.22) $$\Sigma_r(K^{(1)}_{h(n)}) > C > 0 \quad \text{for all } n \in \mathbb{N}$$

whereas

(4.23) $$\Sigma_{r+1}(K^{(1)}_n) \to 0 \quad \text{as} \quad n \to \infty.$$

Let $K^{(1)}_n = E_n^* \operatorname{diag}(\Sigma_1(K^{(1)}_n), \ldots, \Sigma_n(K^{(1)}_n)) F_n$ be the singular value decomposition of $K^{(1)}_n$ which can be chosen in such a way that $P_{n-n_0} E_n^* P_r E_n = 0$ and $P_{n-n_0} F_n^* P_r F_n = 0$. Now write $K^{(1)}_n$ as $K^{(2)}_n + K^{(3)}_n$ with

$$K^{(2)}_n := E_n^* P_r E_n K^{(1)}_n F_n^* P_r F_n.$$

Then one has $P_{n-n_0} K^{(2)}_n = K^{(2)}_n P_{n-n_0} = 0$ for all n, and $\lim_{n\to\infty} \|K^{(3)}_n\| = 0$ due to (4.23).

Consider the sequence $(U_{-h(n)} K_{h(n)} U_{h(n)})_{n\in\mathbb{N}}$. One can assume that this sequence tends *-strongly to a limit operator $\mathbf{K}_h$ of the sequence $\mathbf{K}$ (otherwise pass to a suitable subsequence of h). Then, since P_{n_0} is compact, the sequence $(P_{n_0} J U_{-h(n)} K_{h(n)} U_{h(n)} J P_{n_0})$ tends in the operator norm to $P_{n_0} J \mathbf{K}_h J P_{n_0}$. The entries of that sequence can be written as

$$P_{n_0} J U_{-h(n)} (P_{h(n)} - P_{h(n)-n_0}) K_{h(n)} (P_{h(n)} - P_{h(n)-n_0}) U_{h(n)} J P_{n_0}$$
$$= P_{n_0} J U_{-h(n)} K^{(1)}_{h(n)} U_{h(n)} J P_{n_0} = P_{n_0} J U_{-h(n)} K^{(2)}_{h(n)} U_{h(n)} J P_{n_0} + G_n$$

with a sequence (G_n) tending to zero in the norm. Consequently,

$$P_{n_0} J U_{-h(n)} K^{(2)}_{h(n)} U_{h(n)} J P_{n_0} \to P_{n_0} J \mathbf{K}_h J P_{n_0} \quad \text{in the norm as } n \to \infty.$$

Write the operators on the left hand side as

$$P_{n_0} J U_{-h(n)} K^{(2)}_{h(n)} U_{h(n)} J P_{n_0}$$
$$= P_{n_0} J U_{-h(n)} E^*_{h(n)} P_r E_{h(n)} K^{(2)}_{h(n)} F^*_{h(n)} P_r F_{h(n)} U_{h(n)} J P_{n_0}$$
$$= P_{n_0} J U_{-h(n)} E^*_{h(n)} P_r E_{h(n)} U_{h(n)} J P_{n_0}$$
$$\times P_{n_0} J U_{-h(n)} K^{(2)}_{h(n)} U_{h(n)} J P_{n_0}$$
$$\times P_{n_0} J U_{-h(n)} F^*_{h(n)} P_r F_{h(n)} U_{h(n)} J P_{n_0}$$

(recall that the matrices $E_n^* P_r E_n$ and $F_n^* P_r F_n$ both commute with $P_n - P_{n-n_0}$), and abbreviate the operators

$$P_{n_0} J U_{-h(n)} E^*_{h(n)} P_r E_{h(n)} U_{h(n)} J P_{n_0} \quad \text{and} \quad P_{n_0} J U_{-h(n)} F^*_{h(n)} P_r F_{h(n)} U_{h(n)} J P_{n_0}$$

to $P_n^{E,r}$ and $P_n^{F,r}$, respectively. Each of the operators $P_n^{E,r}$ is an orthogonal projection which acts of the range of P_{n_0}, i.e., on a finite dimensional space with a dimension independent of n. Thus, one can choose a subsequence of $(P_n^{E,r})$ which converges in the norm to an operator $P^{E,r}$. For simplicity we will assume that the sequence $(P_n^{E,r})$ itself enjoys this property. Then $P^{E,r}$ is an orthogonal projection, and being the limit of orthogonal projections of rank r, the rank of $P^{E,r}$ is also r (see, e.g., Propositions 2.2.4 and 2.2.6 in [**52**]). Similarly, one can assume that the sequence $(P_n^{F,r})$ tends in the norm to an orthogonal projection $P^{F,r}$ of rank r. Thus,

$$P_{n_0} J U_{-h(n)} K^{(2)}_{h(n)} U_{h(n)} J P_{n_0} \to P^{E,r} P_{n_0} J \mathbf{K}_h J P_{n_0} P^{F,r} \quad \text{in the norm as } n \to \infty.$$

Moreover, by the definition of $K^{(2)}_{h(n)}$, each operator

$$(4.24) \qquad P_{n_0} J U_{-h(n)} K^{(2)}_{h(n)} U_{h(n)} J P_{n_0} = P_n^{E,r} P_{n_0} J U_{-h(n)} K^{(2)}_{h(n)} U_{h(n)} J P_{n_0} P_n^{F,r}$$

thought of as acting from $\operatorname{im} P_n^{F,r}$ to $\operatorname{im} P_n^{E,r}$ is invertible, and the norm of its inverse is $\Sigma_n(K^{(1)}_{h(n)})^{-1}$ which is less than $1/C$ due to (4.22). Thus, the inverses of the operators (4.24) are uniformly bounded. This implies the norm convergence of these inverses from which we conclude that the norm limit of the operators (4.24), i.e. the operator $P^{E,r} P_{n_0} J \mathbf{K}_h J P_{n_0} P^{F,r}$, is invertible when considered as acting from $\operatorname{im} P^{F,r}$ to $\operatorname{im} P^{E,r}$. Since both ranges have dimension r, this implies that the rank of $\mathbf{K}_h$ is at least r. $\square$

Now we turn over to compact sequences in $\mathcal{S}(\mathbb{Z})$. We consider $\mathcal{S}(\mathbb{Z})$ as a subalgebra of the algebra $\mathcal{F}(\mathbb{Z})$ introduced before Theorem 2.16. In fact we can consider $\mathcal{F}(\mathbb{Z})$ as the algebra which consists of all restrictions of sequences in $\mathcal{F}$ onto the even natural numbers. By restricting the compact sequences in $\mathcal{F}$ to $2\mathbb{N}$ we arrive at the ideal $\mathcal{K}(\mathbb{Z})$ of the *compact sequences* in $\mathcal{F}(\mathbb{Z})$. With these notations, the analog of Theorem 4.17 for compact sequences in $\mathcal{S}(\mathbb{Z})$ reads as follows.

THEOREM 4.22. *The intersection $\mathcal{S}(\mathbb{Z}) \cap \mathcal{K}(\mathbb{Z})$ consists of all sequences $(A_n) \in \mathcal{S}(\mathbb{Z})$ which converge strongly to a compact operator. Equivalently, $\mathcal{S}(\mathbb{Z}) \cap \mathcal{K}(\mathbb{Z})$ is the set of all sequences of the form $(P_n K P_n) + (K_n)$ with K a compact operator and (K_n) a sequence in $\mathcal{K}_0(\mathbb{Z})$.*

The proof is similar to that of Theorem 4.17.

The essential rank of a compact sequence in $\mathcal{K}(\mathbb{Z})$ is – with obvious modifications – defined as the essential rank of a compact sequence in $\mathcal{K}$. In analogy with Theorem 4.18 and formula (4.19) one might conjecture that the essential rank of a sequence in $\mathcal{S}(\mathbb{Z}) \cap \mathcal{K}(\mathbb{Z})$ can be determined by taking a sum of *three* terms, corresponding to the left-hand end, the center, and the right-hand end of the interval $\{-n, -n+1, \ldots, n-1\}$ (or to the points $-1, 0, 1$, respectively, if one prefers to think in terms of Allan/Douglas localization considered in Section 2.8). The following example shows that this conjecture fails.

EXAMPLE. Consider the sequence $(K_n) \in \mathcal{F}(\mathbb{Z})$ defined by

$$K_n := \begin{cases} \operatorname{diag}(1, 0, 0, \ldots, 0) & \text{if } n \text{ is even} \\ \operatorname{diag}(0, \ldots, 0, 0, 1) & \text{if } n \text{ is odd.} \end{cases}$$

This sequence is compact (it consists of matrices of rank one), and it belongs to the algebra $\mathcal{S}(\mathbb{Z})$. Indeed, from Theorem 2.13 (a) one easily concludes that the sequence (L_n) with

$$L_n = \operatorname{diag}(1, 0, 0, \ldots, 0, 0, 1) \in \operatorname{im} R_n$$

belongs to $\mathcal{S}(\mathbb{Z})$, and the same holds evidently for the sequence (D_n) with

$$D_n = \operatorname{diag}(d_{-n}, d_{-n+1}, \ldots, d_{n-1}) \in \operatorname{im} R_n, \qquad d_k := (1 + (-1)^k)/2.$$

Thus, $(K_n) = (L_n)(D_n) \in \mathcal{S}(\mathbb{Z})$. Moreover, the sequence (K_n) has essential rank one. Now observe that the sequence (K_n) can be written as $(PK_n) + (QK_n)$ where both sequences (PK_n) and (QK_n) are of essential rank one themselves. A separate (local) computation of the essential rank at the left-hand as well as at the right-hand end of the interval $\{-n, -n+1, \ldots, n-1\}$ (in a similar way as in Theorem 4.18) would yield the value 1 in both cases. This shows clearly that the essential rank of a compact sequence (K_n) in $\mathcal{S}(\mathbb{Z})$ cannot be determined simply by summing up the local essential ranks. (That there *is* nevertheless a *sum* in identity (4.19) is due to the fact that the behavior of the singular values of the part $P_n K P_n$ of this sum is extremely regular.) □

Thus, in order to get an identity for the essential rank of a compact sequence in $\mathcal{S}(\mathbb{Z})$ in terms of limit operators, one has to guarantee that both ends of the interval $\{-n, -n+1, \ldots, n-1\}$ are treated simultaneously. This can be done by identifying a sequence $(K_n) \in \mathcal{S}(\mathbb{Z})$ with the 2×2-matrix of sequences

$$(4.25) \quad \begin{pmatrix} (PK_nP) & (PK_nQJ) \\ (JQL_nP) & (JQK_nQJ) \end{pmatrix}$$

which belongs to $\mathcal{S}(\mathbb{N})_{2 \times 2}$, and which is in $\mathcal{K}_{2 \times 2}$ if $(K_n) \in \mathcal{K}(\mathbb{Z})$. The essential rank of the sequence (4.25) can be determined by the (evident analogue of) identity (4.19), i.e., as a sum of two terms corresponding to the points 0 and 1.

CHAPTER 5

Fredholmness

5.1. Fredholm sequences

Let again $\mathcal{F}$ be an algebra of matrix sequences with strongly monotonically increasing dimension function. Corresponding to the ideal $\mathcal{K}$ we introduce an appropriate class of Fredholm sequences by calling a sequence $(A_n) \in \mathcal{F}$ Fredholm if it is invertible modulo $\mathcal{K}$. The following properties of Fredholm sequences are obvious.

− Stable sequences are Fredholm.
− Adjoints of Fredholm sequences are Fredholm.
− Products of Fredholm sequences are Fredholm.
− The sum of a Fredholm and a compact sequence is Fredholm.
− The set of all Fredholm sequences is open in $\mathcal{F}$.

For another characterization of Fredholm sequences, let again $0 \leq \sigma_1(A) \leq \ldots \leq \sigma_n(A)$ denote the singular values of an $n \times n$ matrix A.

THEOREM 5.1. *The following conditions are equivalent for a sequence $(A_n) \in \mathcal{F}$:*

(a) The sequence (A_n) is Fredholm.

(b) There is a sequence $(B_n) \in \mathcal{F}$ and a sequence $(J_n) \in \mathcal{K}$ with $\sup_n \operatorname{rank} J_n < \infty$ such that

(5.1) $$B_n A_n = I_n + J_n \quad \text{for all } n \in \mathbb{N}.$$

(c) There is a $k \in \mathbb{N}$ such that

(5.2) $$\liminf_{n \to \infty} \sigma_{k+1}(A_n) > 0.$$

PROOF. $(a) \Rightarrow (b)$: If (A_n) is Fredholm then, by definition, there are sequences $(C_n) \in \mathcal{F}$ and $(K_n) \in \mathcal{K}$ such that

(5.3) $$(C_n)(A_n) = (I_n) + (K_n).$$

Choose a sequence $(L_n) \in \mathcal{K}$ with $\|(L_n) - (K_n)\|_\mathcal{F} < 1/2$ and $\sup \operatorname{rank} L_n < \infty$. Writing (5.3) as

$$(C_n)(A_n) = (I_n) + (K_n - L_n) + (L_n)$$

and taking into account the invertibility of $(I_n) + (K_n - L_n)$ in $\mathcal{F}$, we obtain (5.1) with

$$B_n := (I_n + K_n - L_n)^{-1} C_n \quad \text{and} \quad J_n := (I_n + K_n - L_n)^{-1} L_n.$$

$(b) \Rightarrow (c)$: Let the sequences (B_n) and (J_n) be as in (5.1) and let

$$A_n = E_n^* \Sigma_n F_n \quad \text{with} \quad \Sigma := \operatorname{diag}(\sigma_1(A_n), \ldots, \sigma_{\delta(n)}(A_n))$$

and with unitary matrices E_n, F_n refer to the singular value decomposition of A_n. After multiplication by F_n and F_n^*, the identity (5.1) becomes
$$(F_n B_n E_n^*)(\Sigma_n) = (I_n) + (F_n J_n F_n^*).$$
Abbreviating $C_n := F_n B_n E_n^*$ and $K_n := F_n J_n F_n^*$ we get
$$(5.4) \quad C_n \Sigma_n = C_n \mathrm{diag}\,(\sigma_1(A_n), \ldots, \sigma_{\delta(n)}(A_n)) = I_n + K_n \quad \text{for all } n \in \mathbb{N}$$
where still $\sup \mathrm{rank} K_n < \infty$. We set
$$k := \limsup_{n \to \infty} \mathrm{rank}\, K_n$$
and claim that $\liminf_{n \to \infty} \sigma_{k+1}(A_n) > 0$. Contrary to what we want, assume that this is wrong. Then there exists an infinite subsequence $(n_l)_{l \geq 1}$ of $\mathbb{N}$ such that $\lim_{l \to \infty} \sigma_{k+1}(A_{n_l}) = 0$. Multiplying (5.4) from both sides by P_{k+1}, we get
$$P_{k+1} C_{n_l} \Sigma_{n_l} P_{k+1} = P_{k+1} + P_{k+1} K_{n_l} P_{k+1}.$$
Since
$$\|\Sigma_{n_l} P_{k+1}\| = \|\mathrm{diag}\,(\sigma_1(A_{n_l}), \ldots, \sigma_{k+1}(A_{n_l}), 0, \ldots, 0)\| = \sigma_{k+1}(A_{n_l}) \to 0,$$
one has
$$\lim_{l \to \infty} \|P_{k+1} + P_{k+1} K_{n_l} P_{k+1}\| = 0.$$
Thus, the matrices $P_{k+1} K_{n_l} P_{k+1} \in \mathbb{C}^{(k+1) \times (k+1)}$ are invertible for all sufficiently large n_l. This is impossible since P_{k+1} has rank $k+1$, whereas
$$\mathrm{rank}\, P_{k+1} K_{n_l} P_{k+1} \leq \mathrm{rank}\, K_{n_l} \leq k.$$
This proves the claim which, on its hand, implies assertion (c).

(c) $\Rightarrow$ (a): As in the previous part of the proof, let again $A_n = E_n^* \Sigma_n F_n$ refer to the singular value decomposition of A_n, and let k be a non-negative integer such that
$$\liminf_{n \to \infty} \sigma_{k+1}(A_n) > 0.$$
The choice of k guarantees that the sequence $(\Sigma_n + P_k)_{n \geq 1}$ (with $P_0 := 0$) is stable. Then $(A_n + E_n^* P_k F_n)_{n \in \mathbb{N}}$ is a stable sequence, too. Thus, there are sequences $(C_n) \in \mathcal{F}$ and $(G_n), (H_n) \in \mathcal{G}$ such that
$$(C_n)(A_n + E_n^* P_k F_n) = (I_n) + (G_n) \quad \text{and} \quad (A_n + E_n^* P_k F_n)(C_n) = (I_n) + (H_n),$$
whence
$$(C_n)(A_n) = (I_n) + (G_n) - (C_n E_n^* P_k F_n)$$
and
$$(A_n)(C_n) = (I_n) + (H_n) - (E_n^* P_k F_n C_n).$$
The sequences $(G_n) - (C_n E_n^* P_k F_n)$ and $(H_n) - (E_n^* P_k F_n C_n)$ are of finite essential rank. Hence, (A_n) is invertible modulo $\mathcal{K}$. $\square$

The preceding theorem suggests to introduce the α-*number* $\alpha(\mathbf{A})$ of a Fredholm sequence $\mathbf{A} = (A_n)$ which corresponds to the kernel dimension of a Fredholm operator. By definition, $\alpha(\mathbf{A})$ is the smallest non-negative integer k for which (5.2) is true. Equivalently, $\alpha(\mathbf{A})$ is the smallest non-negative integer k for which there exists a sequence $(B_n) \in \mathcal{F}$ as well as a sequence $(J_n) \in \mathcal{K}$ of essential rank k such that $B_n A_n^* A_n = I_n + J_n$ for all $n \in \mathbb{N}$. The latter fact follows easily from the proof of the preceding theorem.

The *index* of a Fredholm sequence $\mathbf{A}$ is the integer
$$\operatorname{ind}(\mathbf{A}) := \alpha(\mathbf{A}) - \alpha(\mathbf{A}^*).$$
Observe that, in the case at hand, this index is always zero. This is a consequence of the fact that the operators A_n act on finite dimensional spaces which implies that $A_n^* A_n$ and $A_n A_n^*$ have the same eigenvalues even with respect to their multiplicity. So the more interesting quantity associated with a Fredholm sequence seems to be its α-number. On the other hand, the vanishing of the index of (A_n) allows one to make use of the index as a conservation quantity. Some remarkable consequences of this fact were pointed out in [**43**].

The following result provides a relation between the essential norm of a sequence $(A_n) \in \mathcal{F}$ (i.e. the norm of its coset in $\mathcal{F}/\mathcal{K}$) and the singular values $0 \le \Sigma_{\delta(n)}(A_n) \le \ldots \le \Sigma_1(A_n) = \|A_n\|$ of the matrices A_n.

THEOREM 5.2. *For every sequence* $(A_n) \in \mathcal{F}$,
$$\|(A_n) + \mathcal{K}\|_{\mathcal{F}/\mathcal{K}} = \lim_{k \to \infty} \limsup_{n \to \infty} \Sigma_k(A_n). \tag{5.5}$$

Since $k \mapsto \sup_{n \ge k} \Sigma_k(K_n)$ is a monotonically decreasing sequence, the limit can be replaced by an infimum.

PROOF. Consider the singular value decomposition
$$A_n = E_n^* \operatorname{diag}(\Sigma_1(A_n), \ldots, \Sigma_{\delta(n)}(A_n)) F_n$$
of A_n and set, for $n > k > 1$,
$$K_n^{(k)} := E_n^* \operatorname{diag}(\Sigma_1(A_n), \ldots, \Sigma_{k-1}(A_n), 0, \ldots, 0) F_n.$$
Then $\|A_n - K_n^{(k)}\| = \Sigma_k(A_n)$ and, consequently,
$$\|(A_n) - (K_n^{(k)}) + \mathcal{G}\| = \limsup_{n \to \infty} \Sigma_k(A_n).$$
Thus, $\|(A_n) + \mathcal{K}\| \le \limsup_{n \to \infty} \Sigma_k(A_n)$, whence
$$\|(A_n) + \mathcal{K}\| \le \inf_k \limsup_{n \to \infty} \Sigma_k(A_n) = \lim_{k \to \infty} \limsup_{n \to \infty} \Sigma_k(A_n).$$
For the reverse estimate assume that
$$\|(A_n) + \mathcal{K}\| < \inf_k \limsup_{n \to \infty} \Sigma_k(A_n) =: C$$
for a certain sequence $(A_n) \in \mathcal{F}$. Choose a sequence (K_n) of *finite* essential rank such that $\|(A_n) + (K_n)\| < C$. Since $\limsup_{n \to \infty} \Sigma_k(A_n) \ge C$ for each k, we conclude from Lemma 4.4 that
$$\operatorname{ess\,rank}(K_n) \ge k \qquad \text{for each } k \in \mathbb{N}.$$
Thus, (K_n) cannot be of finite essential rank, in contradiction to the choice of (K_n). □

One consequence of the identity (5.5) is the equivalence of (b) and (c) in Theorem 4.5. Another consequence concerns the norm of the essential inverse of a Fredholm sequence.

COROLLARY 5.3. *Let* $(A_n) \in \mathcal{F}$ *be a Fredholm sequence. Then*
$$\|((A_n) + \mathcal{K})^{-1}\|_{\mathcal{F}/\mathcal{K}} = \frac{1}{\sup_k \liminf_{n \to \infty} \sigma_k(A_n)}. \tag{5.6}$$

PROOF. Let l be the α-number of (A_n) and let
$$A_n = E_n^* \text{diag}\,(\sigma_1(A_n), \ldots, \sigma_{\delta(n)}(A_n))F_n$$
be the singular value decomposition of A_n. If $\sigma_{l+1}(A_n) \neq 0$ (which happens for all sufficiently large n), set
$$B_n := E_n^* \text{diag}\,(\sigma_{l+1}(A_n), \ldots, \sigma_{l+1}(A_n), \sigma_{l+2}(A_n), \ldots, \sigma_{\delta(n)}(A_n))F_n$$
with $\sigma_{l+1}(A_n)$ being repeated $l+1$ times. Otherwise, choose B_n to be invertible. Then $(A_n) + \mathcal{K} = (B_n) + \mathcal{K}$, and (B_n) is a stable sequence. In particular,
$$\|((A_n) + \mathcal{K})^{-1}\| = \|((B_n) + \mathcal{K})^{-1}\| = \|(B_n^{-1}) + \mathcal{K}\|.$$
By the preceding theorem,
$$\|((A_n) + \mathcal{K})^{-1}\| = \inf_k \limsup_{n \to \infty} \Sigma_k(B_n^{-1}) = \frac{1}{\sup_k \liminf_{n \to \infty} \sigma_k(B_n)}$$
since $\Sigma_k(B) = \sigma_k(B^{-1})^{-1}$ for each invertible matrix B. The assertion follows since $\sigma_k(B_n) = \sigma_k(A_n)$ for $k > l$. $\square$

5.2. Fredholmness and stability

If A is a Fredholm operator with index 0, then there is an operator K with finite rank such that $A + K$ is invertible. The analog for Fredholm sequences reads as follows. Notice that there is no index obstruction since the index of a Fredholm sequence (of squared matrices) is always 0.

THEOREM 5.4. *If $(A_n) \in \mathcal{F}$ is a Fredholm sequence, then there is a sequence $(K_n) \in \mathcal{K}$ with*
$$\text{ess rank}\,(K_n) \leq \alpha(A_n)$$
such that $(A_n) + (K_n)$ is a stable sequence.

This has been essentially shown in the course of the proof of Corollary 5.3. Let k denote the α-number of (A_n), and let
$$A_n = E_n^* \text{diag}\,(\sigma_1(A_n), \ldots, \sigma_{\delta(n)}(A_n))F_n$$
be the singular value decomposition of A_n. Set
$$K_n := E_n^* \text{diag}\,(\sigma_{k+1}(A_n) - \sigma_1(A_n), \ldots, \sigma_{k+1}(A_n) - \sigma_k(A_n), 0, \ldots, 0)F_n.$$
Then $\text{ess rank}\,(K_n) \leq k$ since each matrix K_n has a rank not greater than k, and the sequence $(A_n) + (K_n)$ is stable.

Notice that this choice of (K_n) implies that
$$\|((A_n) + (K_n) + \mathcal{G})^{-1}\| = \frac{1}{\liminf_{n \to \infty} \sigma_{k+1}(A_n)}.$$

Another immediate consequence of Theorem 5.4 is the following Weyl-type characterization of the *essential spectrum* of a sequence $\mathbf{A} \in \mathcal{F}$, i.e., of the spectrum of the coset $\mathbf{A} + \mathcal{K}$ in the quotient algebra $\mathcal{F}/\mathcal{K}$.

COROLLARY 5.5. *For each sequence $\mathbf{A} \in \mathcal{F}$,*
$$\sigma_{\mathcal{F}/\mathcal{K}}(\mathbf{A} + \mathcal{K}) = \cap_{\mathbf{K} \in \mathcal{K}}\, \sigma_{\mathcal{F}/\mathcal{G}}(\mathbf{A} + \mathbf{K} + \mathcal{G}).$$

Indeed, the inclusion $\subseteq$ holds trivially, and the reverse implication follows from Theorem 5.4.

5.3. Fredholmness and stable regularizability

The relation between Fredholmness and stable regularizability of a sequence in $\mathcal{F}$ differs basically from the well-known relation between Fredholmness and Moore-Penrose invertibility of a bounded linear operator. Indeed, whereas every Moore-Penrose invertible operator is normally solvable, hence Moore-Penrose invertible, the following example shows that a similar relation fails already for sequences in $\mathcal{S}(\mathbb{N})$.

EXAMPLE. Let $(a_n)_{n \geq 1}$ be an enumeration of the rational numbers in the interval $[0, 1/2]$, and set

$$A_n := \begin{pmatrix} a_n & 1 \\ 1 & a_n \end{pmatrix} \quad \text{and} \quad A := \text{diag}\,(A_1, A_2, \ldots) \in L(l^2(\mathbb{N})).$$

The eigenvalues of the matrix A_n are $a_n \pm 1$. Hence, all matrices A_n are invertible, and $\|A_n^{-1}\| \leq 2$. Since

$$P_n A P_n = \begin{cases} \text{diag}\,(A_1, A_2, \ldots, A_k) & \text{if } n = 2k \\ \text{diag}\,(A_1, A_2, \ldots, A_k, a_{k+1}) & \text{if } n = 2k+1, \end{cases}$$

none of the sequences $(P_n A P_n) - \lambda(I_n)$ with $\lambda \in [0, 1/2]$ can be stable. Hence, the interval $[0, 1/4]$ belongs to the spectrum of $\mathbf{A}^*\mathbf{A} + \mathcal{G}$, which shows via Theorem 3.2 that the sequence $\mathbf{A}$ is not stably regularizable. But this sequence is Fredholm. Indeed, let

$$K_n := \begin{cases} \text{diag}\,(0, 0, \ldots, 0) & \text{if } n = 2k \\ \text{diag}\,(0, 0, \ldots, 0, 1) & \text{if } n = 2k+1. \end{cases}$$

The sequence (K_n) is compact (its entries are matrices of rank at most one), and the sequence $(P_n A P_n) + (K_n)$ is stable. Hence, $\mathbf{A}$ is Fredholm. $\square$

Notice that the subsequence $(P_{2n} A P_{2n})_{n \geq 1}$ of $\mathbf{A}$ is stably regularizable (actually, this subsequence is even stable). This is a general phenomenon as the following result shows.

THEOREM 5.6. *Every Fredholm sequence* $\mathbf{A} \in \mathcal{F}$ *possesses a stably regularizable subsequence* $\mathbf{A}_\eta$ *with* $\alpha(\mathbf{A}) = \alpha(\mathbf{A}_\eta)$.

PROOF. Let $\mathbf{A} = (A_n)$ be a Fredholm sequence and $k := \alpha(\mathbf{A})$. Thus,

$$\liminf_{n \to \infty} \sigma_{k+1}(A_n) =: C > 0 \quad \text{and} \quad \liminf_{n \to \infty} \sigma_k(A_n) = 0.$$

Choose a monotonically increasing sequence $\eta : \mathbb{N} \to \mathbb{N}$ such that

$$\lim_{n \to \infty} \sigma_k(A_{\eta(n)}) = 0,$$

and set $\mathbf{A}_\eta = (A_{\eta(n)})_{n \geq 1}$. Then the sequence $\mathbf{A}_\eta$ is Fredholm and it has the same α-number as $\mathbf{A}$. That this sequence is also stably regularizable can be seen as follows. Let

$$A_n = E_n^* \,\text{diag}\,(\sigma_1(A_n), \ldots, \sigma_{\delta(n)}(A_n)) F_n$$

be the singular value decomposition of A_n and define

$$B_n := E_n^* \,\text{diag}\,(0, 0, \ldots, 0, \sigma_{k+1}(A_n), \ldots, \sigma_{\delta(n)}(A_n)) F_n.$$

The sequence $\mathbf{B}_\eta := (B_{\eta(n)})_{n \geq 1}$ is stably regularizable (it is even Moore-Penrose invertible with $\mathbf{B}_\eta^\dagger = (B_{\eta(n)}^\dagger)$ since the matrices $B_{\eta(n)}^\dagger$ are uniformly bounded).

Since the sequences $\mathbf{A}_\eta$ and $\mathbf{B}_\eta$ differ by a sequence tending to zero, the stable regularizability of $\mathbf{A}_\eta$ follows. $\square$

For sequences which are both Fredholm and stably regularizable one has the following quantitative version of the splitting property.

THEOREM 5.7. *Let $\mathbf{A} = (A_n) \in \mathcal{F}$ be a stably regularizable Fredholm sequence, and let $\mathbf{K} + \mathcal{G} \in \mathcal{F}/\mathcal{G}$ denote the Moore-Penrose projection of $\mathbf{A} + \mathcal{G}$. Then*

(a) the number of singular values of A_n which tend to zero (i.e., the number of the singular values of A_n in the block $[0, c_n]$ of the splitting decomposition (3.4)) is not larger than $\alpha(\mathbf{A})$, and there is a subsequence of $\mathbf{A}$ for which this number is exactly $\alpha(\mathbf{A})$.

(b) $\mathbf{K}$ is a compact sequence with essential rank $\alpha(\mathbf{A})$.

This follows easily by working with the singular value decomposition of A_n as in the proof of the previous theorem.

5.4. Fredholmness of the finite sections method

Whereas the stability of the sequence the finite sections method $(P_n A P_n)$ of a general band-dominated operator $A \in \mathcal{A}(\mathbb{N})$ involves a bulk of extra conditions besides the invertibility of A (the uniform invertibility of all limit operators, by Theorem 2.8), the Fredholmness of A implies the Fredholmness of the sequence $(P_n A P_n)$ without any additional ingredients.

THEOREM 5.8. *(a) Let $A \in \mathcal{A}(\mathbb{N})$. The sequence $(P_n A P_n)$ of the finite sections method is Fredholm if and only if the operator A is Fredholm.*

(b) A sequence $\mathbf{A} = (A_n) \in \mathcal{S}(\mathbb{N})$ is Fredholm if and only if its strong limit $W(\mathbf{A})$ is a Fredholm operator.

(c) The mapping $\mathcal{A}(\mathbb{N})/K(l^2(\mathbb{N})) \to \mathcal{S}(\mathbb{N})/(\mathcal{S}(\mathbb{N}) \cap \mathcal{K})$,

$$(5.7) \qquad A + K(l^2(\mathbb{N})) \mapsto (P_n A P_n) + \mathcal{S}(\mathbb{N}) \cap \mathcal{K}$$

is a $*$-isomorphism.

PROOF. (a) Let A be Fredholm, and let R be a regularizer of A, i.e., the operators $K_r := I - RA$ and $K_l := I - AR$ are compact. Then

$$P_n K P_n A P_n = P_n R A P_n - P_n R Q_n A P_n = P_n - P_n K_r P_n - P_n R Q_n A P_n$$

with $Q_n := I - P_n$. Since the compact operator K_r can be approximated as closely as desired by finite rank operators, the sequence $(P_n K_r P_n)$ belongs to the ideal $\mathcal{K}$. We claim that the sequence $(P_n R Q_n A P_n)$ is compact, too. Since A can be approximated as closely as desired by band operators, it is enough to check this claim for a band operator A, and this has already been done in the proof of Theorem 4.17. So we get $(P_n R P_n)(P_n A P_n) - (P_n) \in \mathcal{K}$ and, analogously, $(P_n A P_n)(P_n R P_n) - (P_n) \in \mathcal{K}$. Thus, $(P_n A P_n)$ is a Fredholm sequence.

Let, conversely, $(P_n A P_n)$ be a Fredholm sequence. Then the coset $(P_n A P_n) + \mathcal{K}$ is invertible in $\mathcal{F}/\mathcal{K}$. By inverse closedness of C^*-algebras, this coset is also invertible in $(\mathcal{F}^C + \mathcal{K})/\mathcal{K}$ where $\mathcal{F}^C$ refers to the C^*-subalgebra of $\mathcal{F}$ consisting of all $*$-strongly convergent sequences. The $*$-isomorphy

$$(\mathcal{F}^C + \mathcal{K})/\mathcal{K} \cong \mathcal{F}^C/(\mathcal{F}^C \cap \mathcal{K})$$

implies further that the coset $(P_n A P_n) + (\mathcal{F}^C \cap \mathcal{K})$ is invertible in $\mathcal{F}^C/(\mathcal{F}^C \cap \mathcal{K})$. Thus, there are a sequence $(B_n) \in \mathcal{F}^C$ and sequences $(K_n), (L_n) \in \mathcal{F}^C \cap \mathcal{K}$ with strong limits B, K and L, respectively, such that

$$(B_n)(P_n A P_n) = (P_n) + (K_n) \quad \text{and} \quad (P_n A P_n)(B_n) = (P_n) + (L_n).$$

Passing to the strong limit as $n \to \infty$ yields $BA = I + K$ and $AB = I + L$ with compact operators K and L due to Proposition 4.2. Hence, A is compact.

(b) If (A_n) is a Fredholm sequence, then its strong limit is a Fredholm operator which can be shown as in the previous part of this proof. Conversely, let $W(\mathbf{A})$ be a Fredholm operator for a sequence $\mathbf{A} = (A_n) \in \mathcal{S}(\mathbb{N})$. Then, by part (a), the sequence $(P_n W(\mathbf{A}) P_n)$ is Fredholm, and Theorem 4.17 states that the difference $(P_n W(\mathbf{A}) P_n) - (A_n)$ is a compact sequence. Hence, (A_n) is a Fredholm sequence, too.

(c) Since $(P_n K P_n)$ is a compact sequence for every compact operator K by Lemma 4.20, the mapping (5.7) is correctly defined. The proof of part (a) shows that this mapping is a $*$-homomorphism. Further, the assertion (a) states that this mapping preserves spectra; hence it is an isomorphism. (It is also easy to construct the inverse to the mapping (5.7) via the strong limit W.) $\square$

Notice in this connection that the implication

$$(P_n A P_n) \text{ Fredholm} \implies A \text{ Fredholm}$$

holds for every operator $A \in L(l^2(\mathbb{N}))$.

Our next goal are the α-numbers of the finite sections sequence $(P_n A P_n)$ for Fredholm band-dominated operators A on $l^2(\mathbb{N})$. The following is the main result of this section.

THEOREM 5.9. *Let $A \in l^2(\mathbb{N})$ be a Fredholm band-dominated operator. Then $(P_n A P_n)$ is a Fredholm sequence, and*

(5.8) $\quad \alpha(P_n A P_n) = \dim \ker A + \max \{\dim \ker (Q A_h Q + P) : A_h \in \sigma_+(A)\}.$

PROOF. By Theorem 5.6, there is a subsequence $\mathbf{A}_\eta$ of $\mathbf{A} = (P_n A P_n)$ which is stably regularizable and which has the same α-number as $\mathbf{A}$. For simplicity we assume that $\mathbf{A}_\eta = \mathbf{A}$. Choose a sequence $\mathbf{K} = (K_n)$ with essential rank $\alpha(\mathbf{A})$ such that $\mathbf{K} + \mathcal{G}$ is the Moore-Penrose projection of $\mathbf{A} + \mathcal{G}$ (see Theorem 5.7). Thus, $\mathbf{A}\mathbf{K} \in \mathcal{G}$, and there is a sequence $\mathbf{B} = (B_n) \in \mathcal{S}(\mathbb{N})$ such that $\mathbf{B}(\mathbf{A}^*\mathbf{A} + \mathbf{K}) - (I_n) \in \mathcal{G}$ and $(\mathbf{A}^*\mathbf{A} + \mathbf{K})\mathbf{B} - (I_n) \in \mathcal{G}$. Passing to strong limits with respect to a sequence $h \in \mathcal{H}$ yields that $(Q A_h Q + P)\mathbf{K}_h = 0$ and that the operators $(Q A_h Q + P)^*(Q A_h Q + P) + \mathbf{K}_h$ are invertible. Thus, $\mathbf{K}_h$ is the Moore-Penrose projection of $Q A_h Q + P$ (notice that all operators in $\sigma_{op,1}(\mathbf{K})$ are projections). Since the Moore-Penrose projection of an operator coincides with the orthogonal projection onto its kernel, we get

(5.9) $\quad\quad\quad\quad\quad \operatorname{rank} \mathbf{K}_h = \dim \ker (Q A_h Q + P).$

A similar reasoning shows that the strong limit K of the sequence $\mathbf{K}$ is the Moore-Penrose inverse of A, whence $\operatorname{rank} K = \dim \ker A$. Together with (5.9), this yields the assertion via Theorem 4.18. $\square$

The corresponding result for the Fredholmness of the finite sections sequence of a band-dominated operator on $l^2(\mathbb{Z})$ reads as follows.

THEOREM 5.10. (a) Let $A \in \mathcal{A}(\mathbb{Z})$. The sequence $(R_n A R_n)$ of the finite sections method is Fredholm if and only if the operator A is Fredholm.

(b) A sequence $\mathbf{A} = (A_n) \in \mathcal{S}(\mathbb{Z})$ is Fredholm if and only if its strong limit $W(\mathbf{A})$ is a Fredholm operator.

(c) The mapping
$$\mathcal{A}(\mathbb{Z})/K(l^2(\mathbb{Z})) \to \mathcal{S}(\mathbb{Z})/(\mathcal{S}(\mathbb{Z}) \cap \mathcal{K}(\mathbb{Z})),$$
$$A + K(l^2(\mathbb{Z})) \mapsto (R_n A R_n) + \mathcal{S}(\mathbb{Z}) \cap \mathcal{K}(\mathbb{Z})$$
is a *-isomorphism.

A formula for the α-number of the sequence $(R_n A R_n)$ of a Fredholm band-dominated operator A can be derived in the way sketched at the end of Section 4.5.

5.5. Slowly oscillating coefficients

The formula for the α-number takes a particularly simple form for finite sections of band-dominated operators with slowly oscillating coefficients.

COROLLARY 5.11. Let $A \in l^2(\mathbb{N})$ be a Fredholm band-dominated operator with slowly oscillating coefficients. Then $(P_n A P_n)$ is a Fredholm sequence, and

(5.10) $\quad \alpha(P_n A P_n) = \max\{\dim \ker A, \dim \operatorname{coker} A\}.$

PROOF. Let $A \in \mathcal{A}_{SO}(\mathbb{N})$ be a Fredholm operator. Then each limit operator $A_h \in \sigma_+(A)$ is invertible, and $JQA_h QJ$, considered on $l^2(\mathbb{N})$, is a Fredholm Toeplitz operator. By Coburn's theorem,

$$\dim \ker(JQA_h QJ|_{l^2(\mathbb{N})}) = \dim \ker(QA_h Q + P) = \max\{\operatorname{ind}(QA_h Q + P), 0\}.$$

Since A_h is invertible, one has

$$0 = \operatorname{ind} A = \operatorname{ind}(PA_h P + Q) + \operatorname{ind}(QA_h Q + P),$$

and by Theorem 2.3 (a),

$$\operatorname{ind}(PA_h P + Q) = \operatorname{ind}_+(A_h) = \operatorname{ind}_+(A) = \operatorname{ind}(PAP + Q) = \operatorname{ind} A.$$

Hence,
$$\dim \ker(JQA_h QJ|_{l^2(\mathbb{N})}) = \max\{-\operatorname{ind} A, 0\}.$$

This shows that the kernel dimensions of $JQA_h QJ|_{l^2(\mathbb{N})}$ are independent of the choice of the limit operator A_h. Thus, (5.8) reduces to

$$\alpha(P_n A P_n) = \dim \ker A + \max\{-\operatorname{ind} A, 0\}$$

which gives the assertion. $\square$

There is also a simple direct proof of Corollary 5.11 without invoking (5.8) which runs as follows.

DIRECT PROOF OF COROLLARY 5.11. First let $\operatorname{ind} A =: k \geq 0$. Choose a compact operator $L \in L(l^2(\mathbb{N}))$ such that the operator $V_k A + L$ (which still is an operator with slowly oscillating coefficients) becomes invertible. By Theorem 2.10, the finite sections method $(P_n (V_k A + L) P_n)$ is stable, i.e., there are a sequence $(B_n) \in \mathcal{F}$ as well as a sequence $(G_n) \in \mathcal{G}$ such that

$$(B_n)(P_n(V_k A + L)P_n) = (I_n) + (G_n).$$

Taking into account that $P_n V_k = P_n V_k P_n$ for $k \geq 0$, one gets
$$(B_n)(P_n V_k P_n)(P_n A P_n) = (I_n) - (B_n P_n L P_n) + (G_n),$$
whence $\alpha(P_n A P_n) \leq \operatorname{rank} L$. Since L can be chosen to have a rank which is equal to the kernel dimension of $V_k A$ ([**21**], Chapter 4, Theorem 6.2), and since
$$\ker A \subseteq \ker V_k A \subseteq \ker V_{-k} V_k A \subseteq \ker A,$$
we obtain
(5.11) $$\alpha(P_n A P_n) \leq \dim \ker A.$$

This settles the assertion for operators with non-negative index. In case $\operatorname{ind} A < 0$ one has $\operatorname{ind} A^* > 0$, and (5.11) implies $\alpha(P_n A^* P_n) \leq \dim \ker A^*$. Since the α-number of a sequence of matrices coincides with the α-number of its adjoint sequence, and since $\dim \ker A^* = \dim \operatorname{coker} A$, we arrive at
(5.12) $$\alpha(P_n A P_n) \leq \max\{\dim \ker A, \dim \operatorname{coker} A\}$$
holding independently of the sign of the index of A.

For the reverse inequality, let $l := \alpha(P_n A P_n)$ and choose a sequence $(B_n) \in \mathcal{F}$ and a sequence $(J_n) \in \mathcal{K}$ of essential rank l such that $B_n P_n A P_n = P_n + J_n$. The sequential compactness of the unit ball in $L(l^2(\mathbb{N}))$ allows one to choose weakly convergent subsequences (B_{n_k}) of (B_n) and (J_{n_k}) of (J_n) with limits B and J, respectively. Recall further that the product of a weakly convergent sequence with limit C and a *-strongly convergent sequence with limit D is weakly convergent with limit CD. Thus, passing to the weak limit as $k \to \infty$ in $B_{n_k} P_{n_k} A P_{n_k} = P_{n_k} + J_{n_k}$ yields $BA = I + J$ with $\operatorname{rank} J \leq l$ by Proposition 4.2 (a). If now $x \in \ker A$, then $x + Jx = BAx = 0$, whence $\ker A \subseteq \operatorname{im} J$. Thus,
$$\dim \ker A \leq \dim \operatorname{im} J = \operatorname{rank} J \leq l.$$
Since $\alpha(P_n A^* P_n) = \alpha(P_n A P_n)$, we also have
$$\dim \operatorname{coker} A = \dim \ker A^* \leq l,$$
which finally gives $\max\{\dim \ker A, \dim \operatorname{coker} A\} \leq \alpha(P_n A P_n)$. Together with (5.12), this implies the assertion. $\square$

COROLLARY 5.12. *Let* $a \in C(\mathbb{T})$ *be a continuous function without zeros on the unit circle* $\mathbb{T}$, *and let* $K \in L(l^2(\mathbb{N}))$ *be compact. Then* $(P_n(T(a) + K)P_n)$ *is a Fredholm sequence, and*
(5.13) $$\alpha(P_n(T(a) + K)P_n) = \dim \ker(T(a) + K) + \dim \ker T(\tilde{a})$$
with $\tilde{a}(t) := a(t^{-1})$.

PROOF. One easily checks that $T(\tilde{a}) = CT(a)^*C$ where $C : l^2(\mathbb{N}) \to l^2(\mathbb{N})$ is the ($\mathbb{R}$-linear) operator of conjugation $(x_n) \mapsto (\overline{x_n})$. Thus,
(5.14) $$\dim \ker T(\tilde{a}) = \dim \operatorname{coker} T(a) \quad \text{and} \quad \dim \operatorname{coker} T(\tilde{a}) = \dim \ker T(a)$$
which holds for arbitrary functions $a \in L^\infty(\mathbb{T})$ such that $T(a)$ is Fredholm. Let now a be a continuous function without zeros on $\mathbb{T}$ and let K be compact. In case $\operatorname{ind}(T(a) + K) \geq 0$ we have $\operatorname{ind} T(a) \geq 0$, and Coburn's theorem implies that $\dim \operatorname{coker} T(a) = \dim \ker T(\tilde{a}) = 0$. Thus, in this case, (5.10) immediately implies
$$\alpha(P_n(T(a) + K)P_n) = \dim \ker(T(a) + K)$$
$$= \dim \ker(T(a) + K) + \dim \ker T(\tilde{a}).$$

Now assume $\operatorname{ind}(T(a) + K) < 0$. As above, we conclude from Coburn's theorem that $\dim \operatorname{coker} T(\tilde{a}) = 0$. Further, we know from (5.14) that

$$\dim \ker (T(a) + K) - \dim \operatorname{coker} (T(a) + K)$$
$$= -\dim \ker T(\tilde{a}) + \dim \operatorname{coker} T(\tilde{a}) = -\dim \ker T(\tilde{a}),$$

whence via (5.10)

$$\alpha(P_n(T(a) + K)P_n) = \dim \operatorname{coker} (T(a) + K)$$
$$= \dim \ker (T(a) + K) + \dim \ker T(\tilde{a})$$

also in this case. $\square$

The identity (5.13) is well-known. In fact, this identity was the initial observation which eventually led to the discovery of a Fredholm theory for approximation sequences in [**43, 42, 47, 23**].

CHAPTER 6

Essential fractality

Roughly speaking, a C^*-subalgebra $\mathcal{B}$ of the algebra $\mathcal{F}$ is said to be fractal if each sequence $(A_n) \in \mathcal{B}$ can be rediscovered from each (infinite) subsequence of (A_n) modulo a sequence in $\mathcal{G}$. Sequences in fractal algebras exhibit excellent asymptotic properties. For example, we already mentioned that
$$\limsup_{n\to\infty} \sigma(A_n^* A_n) = \sigma(\mathbf{A}^*\mathbf{A} + \mathcal{G})$$
for every sequence $\mathbf{A} = (A_n) \in \mathcal{F}$. If the sequence (A_n) belongs to a fractal subalgebra of $\mathcal{F}$, then the limes superior can be replaced by the Hausdorff limit.

It turns out that the algebras of the finite sections method for band-dominated operators are not fractal (a reasoning will be given later on), but these algebras enjoy a weaker form of fractality which we call essential fractality and which will be discussed briefly in this section.

6.1. Fractality of quotient maps

Let $\mathcal{F} = \mathcal{F}^\delta$ be an algebra of matrix sequences with dimension function δ and $\mathcal{G} = \mathcal{G}^\delta$ its associated ideal of zero sequences, and let $\eta : \mathbb{N} \to \mathbb{N}$ be a strongly monotonically increasing sequence. We denote by $\mathcal{F}_\eta = \mathcal{F}_\eta^\delta$ the algebra of matrix sequences with dimension function $\delta \circ \eta$, and we write $\mathcal{G}_\eta = \mathcal{G}_\eta^\delta$ for its associated ideal of zero sequences. Thus, $\mathcal{F}_\eta^\delta = \mathcal{F}^{\delta \circ \eta}$. Further, we let R_η stand for the restriction mapping
$$R_\eta : \mathcal{F} \to \mathcal{F}_\eta, \qquad (A_n) \mapsto (A_{\eta(n)}).$$
The mapping R_η is a *-homomorphism from $\mathcal{F}$ onto $\mathcal{F}_\eta$ which moreover maps $\mathcal{G}$ onto $\mathcal{G}_\eta$. For each C^*-subalgebra $\mathcal{B}$ of $\mathcal{F}$, we write $\mathcal{B}_\eta$ for the image of $\mathcal{B}$ in $\mathcal{F}_\eta$ under the mapping R_η, which is a C^*-subalgebra of $\mathcal{F}_\eta$.

DEFINITION 6.1. Let $\mathcal{B}$ be a C^*-subalgebra of $\mathcal{F}$. A *-homomorphism W of $\mathcal{B}$ into a C^*-algebra $\mathcal{C}$ is called *fractal* if, for every strongly monotonically increasing sequence $\eta : \mathbb{N} \to \mathbb{N}$, there is a *-homomorphism $W_\eta : \mathcal{B}_\eta \to \mathcal{C}$ such that
$$W = W_\eta R_\eta|_\mathcal{B}.$$

Thus, for each sequence $\mathbf{A} \in \mathcal{B}$, it is possible to reconstruct the image $W(\mathbf{A})$ under a fractal homomorphism W from each subsequence $\mathbf{A}_\eta$ of $\mathbf{A}$.

THEOREM 6.2. *Let $\mathcal{J}$ be a closed ideal of $\mathcal{F}$ and let $\mathcal{B}$ be a C^*-subalgebra of $\mathcal{F}$. The canonical homomorphism*
$$\pi^\mathcal{J} : \mathcal{B} \to \mathcal{B}/(\mathcal{B} \cap \mathcal{J})$$
is fractal if and only if the following implication holds for each sequence $\mathbf{A} \in \mathcal{B}$ and for each strongly monotonically increasing sequence $\eta : \mathbb{N} \to \mathbb{N}$:

(6.1) $$R_\eta(\mathbf{A}) \in (\mathcal{B} \cap \mathcal{J})_\eta \implies \mathbf{A} \in \mathcal{J}.$$

PROOF. Let $\pi^{\mathcal{J}}$ be a fractal homomorphism, and let $R_\eta(\mathbf{A}) \in (\mathcal{B} \cap \mathcal{J})_\eta$ for a sequence $\mathbf{A} \in \mathcal{B}$. We choose a sequence $\mathbf{J} \in \mathcal{B} \cap \mathcal{J}$ such that $R_\eta(\mathbf{A}) = R_\eta(\mathbf{J})$. Applying the homomorphism $\pi_\eta^{\mathcal{J}}$ to both sides of this equality we obtain $\pi^{\mathcal{J}}(\mathbf{A}) = \pi^{\mathcal{J}}(\mathbf{J}) = 0$, whence $\mathbf{A} \in \mathcal{J}$.

For the reverse implication, let $\mathbf{A}$ and $\mathbf{B}$ be sequences in $\mathcal{B}$ with $R_\eta(\mathbf{A}) = R_\eta(\mathbf{B})$. Then $R_\eta(\mathbf{A} - \mathbf{B}) = 0 \in (\mathcal{B} \cap \mathcal{J})_\eta$, and (6.1) implies that $\mathbf{A} - \mathbf{B} \in \mathcal{B} \cap \mathcal{J}$. Thus, the mapping
$$\pi_\eta^{\mathcal{J}} : \mathcal{B}_\eta \to \mathcal{B}/(\mathcal{B} \cap \mathcal{J}), \qquad R_\eta(\mathcal{A}) \mapsto \mathbf{A} + \mathcal{J}$$
is correctly defined, and one obviously has $\pi_\eta^{\mathcal{J}} R_\eta|_\mathcal{B} = \pi^{\mathcal{J}}$. □

In case $\mathcal{J} = \mathcal{G}$, the implication (6.1) can be replaced be a stronger one.

THEOREM 6.3. *Let $\mathcal{B}$ be a C^*-subalgebra of $\mathcal{F}$. The canonical homomorphism*
$$\pi^{\mathcal{G}} : \mathcal{B} \to \mathcal{B}/(\mathcal{B} \cap \mathcal{G})$$
is fractal if and only if the following implication holds for each sequence $\mathbf{A} \in \mathcal{B}$ and for each strongly monotonically increasing sequence $\eta : \mathbb{N} \to \mathbb{N}$:

(6.2) $\qquad\qquad R_\eta(\mathbf{A}) \in \mathcal{G}_\eta \quad \Longrightarrow \quad \mathbf{A} \in \mathcal{G}.$

PROOF. We have only to show that (6.2) implies the fractality of $\pi^{\mathcal{G}}$. Contrary to what we want to show, assume there are a strongly monotonically increasing sequence η and a sequence $\mathbf{A} = (A_n) \in \mathcal{B}$ with $R_\eta(\mathbf{A}) \in \mathcal{G}_\eta$, for which (A_n) is not in $\mathcal{G}$, i.e.
$$\|\pi^{\mathcal{G}}(\mathbf{A})\| = \|(A_n) + \mathcal{G}\| = \limsup \|A_n\| \geq C > 0$$
with some constant C. Since $R_\eta(\mathbf{A}) = (A_{\eta(n)}) \in \mathcal{G}_\eta$, there is an n_0 such that $\|A_{\eta(n)}\| \leq C/2$ for $n \geq n_0$. Define a monotonically increasing sequence $\mu : \mathbb{N} \to \mathbb{N}$ by $\mu(n) := \eta(n + n_0)$. Then $\|R_\mu(\mathbf{A})\| \leq C/2$. On the other hand, the fractality of $\pi^{\mathcal{G}}$ entails that
$$C \leq \|\pi_\mu^{\mathcal{G}} R_\mu(\mathbf{A})\| \leq \|\pi_\mu^{\mathcal{G}}\| \, \|R_\mu(\mathbf{A})\| = \|R_\mu(\mathbf{A})\|,$$
which is a contradiction. Hence, $\mathbf{A} \in \mathcal{G}$. □

COROLLARY 6.4. *Let $\mathcal{B}$ be a C^*-subalgebra of $\mathcal{F}$ for which the canonical homomorphism $\pi^{\mathcal{G}} : \mathcal{B} \to \mathcal{B}/(\mathcal{B} \cap \mathcal{G})$ is fractal. Then*

(6.3) $\qquad\qquad (\mathcal{B} \cap \mathcal{G})_\eta = \mathcal{B}_\eta \cap \mathcal{G}_\eta$

for each strongly monotonically increasing sequence $\eta : \mathbb{N} \to \mathbb{N}$.

PROOF. The implication $\subseteq$ in (6.3) is obvious, and the reverse implication follows from Theorem 6.3. □

Neither the implication (6.2) nor its corollary (6.3) remain valid for arbitrary ideals $\mathcal{J}$ of $\mathcal{F}$ in place of $\mathcal{G}$ as the following example shows.

EXAMPLE. Let $\mathcal{B} = \mathcal{S}_{\mathbb{C}}(\mathbb{N})$ be the algebra of the finite sections method for Toeplitz operators with continuous generating function. Further we choose
$$\mathcal{J} := \{(K_n) \in \mathcal{K} : \lim \|K_{2n}\| = 0\}.$$
Then $\mathcal{J}$ is a closed ideal of $\mathcal{F}$, and

(6.4) $\qquad\qquad \mathcal{S}_{\mathbb{C}}(\mathbb{N}) \cap \mathcal{J} = \mathcal{G}.$

Indeed, the inclusion $\mathcal{G} \subseteq \mathcal{S}_\mathbb{C}(\mathbb{N}) \cap \mathcal{J}$ is evident. Let, conversely, $(A_n) \in \mathcal{S}_\mathbb{C}(\mathbb{N}) \cap \mathcal{J}$. By Theorem 2.13 (a), there is a unique representation
$$A_n = P_n T(a) P_n + P_n K P_n + W_n L W_n + G_n$$
with a continuous function a, compact operators K and L and a sequence $(G_n) \in \mathcal{G}$. This representation implies that the strong limits
$$\text{s-lim } A_n P_n = T(a) + K \quad \text{and} \quad \text{s-lim } W_n A_n W_n = T(\tilde{a}) + L$$
with $\tilde{a}(t) = a(1/t)$ exist. Since $(A_n) \in \mathcal{J}$, one gets $T(a) + K = T(\tilde{a}) + L = 0$, whence $a = 0$ and $K = L = 0$. Thus, $(A_n) \in \mathcal{G}$.

As a consequence of (6.4), the canonical homomorphism
$$\pi^\mathcal{J} : \mathcal{S}_\mathbb{C}(\mathbb{N}) \to \mathcal{S}_\mathbb{C}(\mathbb{N})/(\mathcal{S}_\mathbb{C}(\mathbb{N}) \cap \mathcal{J}) = \mathcal{S}_\mathbb{C}(\mathbb{N})/\mathcal{G}$$
is fractal. Actually, it coincides with the homomorphism $\pi^\mathcal{G}$ which is fractal as one easily checks be employing the strong limits considered above (a detailed proof is in [23], Corollary 1.70, for example). On the other hand, the inclusion $(\mathcal{S}_\mathbb{C}(\mathbb{N}) \cap \mathcal{J})_\eta \subset \mathcal{S}_\mathbb{C}(\mathbb{N})_\eta \cap \mathcal{J}_\eta$ is proper for the sequence $\eta(n) := 2n+1$. The point is that by (6.4), one has $(\mathcal{S}(C) \cap \mathcal{J})_\eta = \mathcal{G}_\eta$, whereas the sequence $(P_{2n+1} K P_{2n+1})$ belongs to $\mathcal{S}(C)_\eta \cap \mathcal{J}_\eta$ for each compact operator K. $\square$

6.2. $\mathcal{J}$-fractal algebras

The considerations in the previous subsection suggest the following definition.

DEFINITION 6.5. Let $\mathcal{J}$ be a closed ideal of $\mathcal{F}$. A C^*-subalgebra $\mathcal{B}$ of $\mathcal{F}$ is called $\mathcal{J}$-fractal if it owns the following properties:
(a) the canonical homomorphism $\pi^\mathcal{J} : \mathcal{B} \to \mathcal{B}/(\mathcal{B} \cap \mathcal{J})$ is fractal, and
(b) $(\mathcal{B} \cap \mathcal{J})_\eta = \mathcal{B}_\eta \cap \mathcal{J}_\eta$ for each strongly monotonically increasing sequence $\eta : \mathbb{N} \to \mathbb{N}$.

The following results show that $\mathcal{J}$-fractality implies what one expects: A sequence in a $\mathcal{J}$-fractal algebra belongs to $\mathcal{J}$ or is invertible modulo $\mathcal{J}$ if and only if at least one of its subsequences has this property.

THEOREM 6.6. Let $\mathcal{J}$ be a closed ideal of $\mathcal{F}$. A C^*-subalgebra $\mathcal{B}$ of $\mathcal{F}$ is $\mathcal{J}$-fractal if and only if the following implication holds for each sequence $\mathbf{A} \in \mathcal{B}$ and for each strongly monotonically increasing sequence $\eta : \mathbb{N} \to \mathbb{N}$:
(6.5) $$R_\eta(\mathbf{A}) \in \mathcal{J}_\eta \implies \mathbf{A} \in \mathcal{J}.$$

PROOF. Let $\mathcal{B}$ be $\mathcal{J}$-fractal, and let $\mathbf{A} \in \mathcal{B}$ be a sequence with $R_\eta(\mathbf{A}) \in \mathcal{J}_\eta$. Then, by property (b) in Definition 6.5,
$$R_\eta(\mathbf{A}) \in \mathcal{B}_\eta \cap \mathcal{J}_\eta = (\mathcal{B} \cap \mathcal{J})_\eta,$$
and property (a) of that definition implies via Theorem 6.2 that $\mathbf{A} \in \mathcal{J}$.

Conversely, let the implication (6.5) hold for each strongly monotonically increasing sequence $\eta : \mathbb{N} \to \mathbb{N}$. As in the proof of Theorem 6.2, one gets the fractality of the canonical homomorphism $\pi^\mathcal{J}$, which is property (a) in Definition 6.5. For property (b), note that the inclusion $\subseteq$ is obvious. For the reverse inclusion, let $\mathbf{A}$ be a sequence in $\mathcal{F}$ with $R_\eta(\mathbf{A}) \in \mathcal{B}_\eta \cap \mathcal{J}_\eta$. Then there are sequences $\mathbf{B} \in \mathcal{B}$ and $\mathbf{J} \in \mathcal{J}$ such that $R_\eta(\mathbf{A}) = R_\eta(\mathbf{B}) = R_\eta(\mathbf{J})$. Since $R_\eta(\mathbf{B}) \in \mathcal{J}_\eta$, the implication (6.5) gives $\mathbf{B} \in \mathcal{J}$. Hence, $R_\eta(\mathbf{B}) \in (\mathcal{B} \cap \mathcal{J})_\eta$. Since $R_\eta(\mathbf{B}) = R_\eta(\mathbf{A})$, one also has $R_\eta(\mathbf{A}) \in (\mathcal{B} \cap \mathcal{J})_\eta$. $\square$

THEOREM 6.7. *Let $\mathcal{J}$ be a closed ideal of $\mathcal{F}$ and $\mathcal{B}$ a $\mathcal{J}$-fractal C^*-subalgebra of $\mathcal{F}$ which contains the identity element. Then the following implication holds for each sequence $\mathbf{A} \in \mathcal{B}$ and for each strongly monotonically increasing sequence $\eta : \mathbb{N} \to \mathbb{N}$:*

(6.6) $\quad R_\eta(\mathbf{A}) + \mathcal{J}_\eta$ *invertible in* $\mathcal{F}_\eta / \mathcal{J}_\eta \implies \mathbf{A} + \mathcal{J}$ *invertible in* $\mathcal{F}/\mathcal{J}$.

PROOF. Let $\mathbf{A}$ be a sequence in $\mathcal{B}$ for which the coset $R_\eta(\mathbf{A}) + \mathcal{J}_\eta$ is invertible in $\mathcal{F}_\eta/\mathcal{J}_\eta$. Due to the inverse closedness of C^*-algebras, this coset is also invertible in $(\mathcal{B}_\eta + \mathcal{J}_\eta)/\mathcal{J}_\eta$. Since the latter algebra is *-isomorphic to $\mathcal{B}_\eta/(\mathcal{B}_\eta \cap \mathcal{J}_\eta)$ and by property (b) of $\mathcal{J}$-fractality, the coset $R_\eta(\mathbf{A}) + (\mathcal{B} \cap \mathcal{J})_\eta$ is invertible in $\mathcal{B}_\eta/(\mathcal{B} \cap \mathcal{J})_\eta$. Choose sequences $\mathbf{B} \in \mathcal{B}$ and $\mathbf{J} \in \mathcal{B} \cap \mathcal{J}$ such that
$$R_\eta(\mathbf{A})\, R_\eta(\mathbf{B}) = R_\eta(\mathbf{I}) + R_\eta(\mathbf{J})$$
where $\mathbf{I} = (I_n)$ refers to the identity element of $\mathcal{F}$. Applying the homomorphism $\pi_\eta^\mathcal{J}$ to both sides of this equality one gets
$$\pi^\mathcal{J}(\mathbf{A})\, \pi^\mathcal{J}(\mathbf{B}) = \pi^\mathcal{J}(\mathbf{I}) + \pi^\mathcal{J}(\mathbf{J})$$
which shows that $\mathbf{AB} - \mathbf{I} \in \mathcal{J}$. Hence, $\mathbf{A}$ is invertible modulo $\mathcal{J}$ from the right-hand side, and its invertibility from the left-hand side follows analogously. $\square$

To summarize these results we agree upon saying that the subsequence $R_\eta(\mathbf{A})$ of a sequence $\mathbf{A} \in \mathcal{F}$ *belongs to the ideal $\mathcal{J}$* or is *invertible modulo $\mathcal{J}$* if $R_\eta(\mathbf{A}) \in \mathcal{J}_\eta$ or if the coset $R_\eta(\mathbf{A}) + \mathcal{J}_\eta$ is invertible in $\mathcal{F}_\eta/\mathcal{J}_\eta$.

COROLLARY 6.8. *Let $\mathcal{J}$ be a closed ideal of $\mathcal{F}$ and $\mathcal{B}$ a $\mathcal{J}$-fractal C^*-subalgebra of $\mathcal{F}$ which contains the identity element. Then*

(a) a sequence $\mathbf{A} \in \mathcal{B}$ belongs to $\mathcal{J}$ if and only if one of the subsequences of $\mathbf{A}$ belongs to $\mathcal{J}$.

(b) a sequence $\mathbf{A} \in \mathcal{B}$ is invertible modulo $\mathcal{J}$ if and only if one of the subsequences of $\mathbf{A}$ is invertible modulo $\mathcal{J}$.

One feature of sequences in $\mathcal{J}$-fractal algebras is that quotient norms are preserved when passing to subsequences.

THEOREM 6.9. *Let $\mathcal{J}$ be a closed ideal of $\mathcal{F}$, $\mathcal{B}$ a $\mathcal{J}$-fractal C^*-subalgebra of $\mathcal{F}$, and $\eta : \mathbb{N} \to \mathbb{N}$ a strongly monotonically increasing sequence. Then*
$$\|R_\eta(\mathbf{A}) + \mathcal{J}_\eta\| = \|\mathbf{A} + \mathcal{J}\|$$
for each sequence $\mathcal{A} \in \mathcal{B}$.

PROOF. For arbitrary sequences $\mathbf{A} \in \mathcal{B}$ and $\mathbf{J} \in \mathcal{J}$,
$$\|R_\eta(\mathbf{A}) + \mathcal{J}_\eta\| \leq \|R_\eta(\mathbf{A}) + R_\eta(\mathbf{J})\| \leq \|\mathbf{A} + \mathbf{J}\|.$$
Taking the infimum over all sequences $\mathbf{J} \in \mathcal{J}$ on the right-hand side yields
$$\|R_\eta(\mathbf{A}) + \mathcal{J}_\eta\| \leq \|\mathbf{A} + \mathcal{J}\|.$$
The reverse estimate follows from
$$\begin{aligned}\|\mathbf{A} + \mathcal{J}\| &= \|\mathbf{A} + \mathbf{J} + \mathcal{J}\| = \|\pi^\mathcal{J}(\mathbf{A} + \mathbf{J})\| \\ &= \|\pi_\eta^\mathcal{J} R_\eta(\mathbf{A} + \mathbf{J})\| \leq \|R_\eta(\mathbf{A}) + R_\eta(\mathbf{J})\|\end{aligned}$$
(which holds for arbitrary sequences $\mathbf{J} \in \mathcal{J}$) by taking again the infimum over all sequences $\mathbf{J} \in \mathcal{J}$ on the right-hand side. $\square$

Notice that we have only used the fractality of the homomorphism $\pi^{\mathcal{J}}$, not the fractality of the algebra itself. We conclude with some technical results.

PROPOSITION 6.10. *Let $\mathcal{J}$ be a closed ideal of $\mathcal{F}$ and $\mathcal{B}$ be a $\mathcal{J}$-fractal C^*-subalgebra of $\mathcal{F}$. Then*

(a) *each C^*-subalgebra of $\mathcal{B}$ is $\mathcal{J}$-fractal.*
(b) *if $\mathcal{I}$ is an ideal of $\mathcal{F}$ with $\mathcal{J} \subseteq \mathcal{I}$ and if $(\mathcal{B} \cap \mathcal{I})_\eta = \mathcal{B}_\eta \cap \mathcal{I}_\eta$ for each strongly monotonically increasing sequence η, then $\mathcal{B}$ is $\mathcal{I}$-fractal.*

PROOF. (a) Let $\mathcal{C}$ be a C^*-subalgebra of $\mathcal{B}$, and let $\mathbf{A} \in \mathcal{C}$ and $R_\eta(\mathbf{A}) \in \mathcal{C}_\eta \cap \mathcal{J}_\eta$ for a certain strongly monotonically increasing sequence η. Then $R_\eta(\mathbf{A}) \in \mathcal{B}_\eta \cap \mathcal{J}_\eta$ and, since $\mathcal{B}$ is $\mathcal{J}$-fractal, $\mathbf{A} \in \mathcal{J}$ by Theorem 6.6. The same theorem applies once more to state the $\mathcal{J}$-fractality of $\mathcal{C}$.

(b) Let $R_\eta(\mathbf{A}) \in \mathcal{I}_\eta$ for a sequence $\mathbf{A} \in \mathcal{B}$ and a strongly monotonically increasing sequence η. Then, by hypothesis, $R_\eta(\mathbf{A}) \in (\mathcal{B} \cap \mathcal{I})_\eta$. Choose a sequence $\mathbf{J} \in \mathcal{B} \cap \mathcal{I}$ with $R_\eta(\mathbf{A}) = R_\eta(\mathbf{J})$. The $\mathcal{J}$-fractality of $\mathcal{B}$ implies that $\mathbf{A} - \mathbf{J} \in \mathcal{J}$, whence $\mathbf{A} \in \mathcal{I} + \mathbf{J} \subseteq \mathcal{I}$. By Theorem 6.6, $\mathcal{B}$ is $\mathcal{I}$-fractal. □

6.3. Essential fractality and singular values

Let now $\mathcal{F}$ be an algebra of matrix sequences with dimension function δ. In this setting we will mainly encounter two instances of $\mathcal{J}$-fractal algebras: those with $\mathcal{J} = \mathcal{G}$, the ideal of the zero sequences, and those with $\mathcal{J} = \mathcal{K}$, the ideal of the compact sequences. In the first case we know from Corollary 6.4 that an algebra $\mathcal{B}$ is $\mathcal{G}$-fractal if and only if the canonical homomorphism $\pi^\mathcal{G}$ is fractal, which is exactly the definition of the fractality of an algebra given in [**23**], Section 1.6. So we will call $\mathcal{G}$-fractal algebras simply *fractal* in what follows, whereas we call $\mathcal{K}$-fractal algebras *essentially fractal*. In these terms, Corollary 6.8 reads as follows.

COROLLARY 6.11. *Let $\mathcal{B}$ be a fractal C^*-subalgebra of $\mathcal{F}$ which contains the identity element. Then a sequence $\mathbf{A} \in \mathcal{B}$ tends to zero (resp. is stable) and only if one of the subsequences of $\mathbf{A}$ tends to zero (resp. is stable).*

COROLLARY 6.12. *Let $\mathcal{B}$ be an essentially fractal C^*-subalgebra of $\mathcal{F}$ which contains the identity element. Then a sequence $\mathbf{A} \in \mathcal{B}$ is compact (resp. Fredholm) and only if one of the subsequences of $\mathbf{A}$ is compact (resp. Fredholm).*

The essential fractality has some striking consequences for the behavior of the smallest singular values.

THEOREM 6.13. *Let $\mathcal{B}$ be an essentially fractal C^*-subalgebra of $\mathcal{F}$ which contains the identity sequence. Then a sequence $(A_n) \in \mathcal{B}$ is Fredholm if and only if there is a $k \in \mathbb{N}$ such that*

(6.7) $$\limsup_{n \to \infty} \sigma_k(A_n) > 0.$$

PROOF. If (A_n) is Fredholm then, by Theorem 5.1, $\liminf \sigma_k(A_n) > 0$ for some k, whence (6.7). Conversely, let (6.7) hold for some k. We choose a strongly monotonically increasing sequence $\eta : \mathbb{N} \to \mathbb{N}$ such that

$$\lim_{n \to \infty} \sigma_k(A_{\eta(n)}) > 0.$$

By Theorem 5.1, the sequence $(A_{\eta(n)})_{n \geq 1}$ is Fredholm, and Corollary 6.12 implies the Fredholmness of the sequence (A_n). □

Consequently, if a sequence (A_n) in an essentially fractal C^*-subalgebra of $\mathcal{F}$ is *not* Fredholm, then
$$\lim_{n \to \infty} \sigma_k(A_n) = 0 \quad \text{for each } k \in \mathbb{N}.$$
In analogy with operator theory one might call a sequence with this property *not normally solvable*.

COROLLARY 6.14. *Let $\mathcal{B}$ be an essentially fractal C^*-subalgebra of $\mathcal{F}$ which contains the identity sequence. Then a sequence $(A_n) \in \mathcal{B}$ is either Fredholm or not normally solvable.*

6.4. Essential fractality of the finite sections algebras

Here we consider the fractality properties of the finite sections algebras $\mathcal{S}(\mathbb{N})$ and $\mathcal{S}(\mathbb{Z})$. We have already observed that the subalgebra $\mathcal{S}_{\mathbb{C}}(\mathbb{N})$ of $\mathcal{S}(\mathbb{N})$ of the finite sections method for Toeplitz operators is fractal, and the same is true for the subalgebra $\mathcal{S}_{PC}(\mathbb{Z})$ of $\mathcal{S}(\mathbb{Z})$ which is generated by the finite sections sequences $(R_n L(a) R_n)$ of Laurent operators with continuous generating function and by the sequence $(R_n P R_n)$. Due to the possible much more irregular behavior of the coefficients of the operators (being merely bounded instead of constant), one cannot expect the same excellent fractality properties for the finite sections algebras $\mathcal{S}(\mathbb{N})$ and $\mathcal{S}(\mathbb{Z})$. Consider, for example, the finite sections of the (unitary) band operator
$$A := \text{diag}\left(\begin{pmatrix} 0 & 1 \\ 1 & 0 \end{pmatrix}, \begin{pmatrix} 0 & 1 \\ 1 & 0 \end{pmatrix}, \ldots\right).$$
Then $(P_{2n} A P_{2n})$ is a stable sequence (of unitary matrices), but no entry of the sequence $(P_{2n+1} A P_{2n+1})$ is invertible. By Corollary 6.11, neither the algebra $\mathcal{S}(\mathbb{N})$ nor any other subalgebra of $\mathcal{S}(\mathbb{N})$ which contains the sequence $(P_n A P_n)$ is fractal. Also the algebra $\mathcal{S}_{SO}(\mathbb{N})$ of the finite sections method for band-dominated operators with slowly oscillating coefficients fails to be fractal. To see this, choose a function $a \in SO(\mathbb{N})$ which oscillates between 0 and 1 (for example, take $a(n) := |\sin \sqrt{n}|$). The sequence $(W_n P_1 W_n)$ belongs to $\mathcal{S}_{SO}(\mathbb{N})$ (it even belongs to $\mathcal{S}_{\mathbb{C}}(\mathbb{N})$ by Theorem 2.13), and so does the sequence $(A_n) := (P_n a P_n)(W_n P_1 W_n)$ where
$$A_n = \text{diag}\,(0, 0, \ldots, 0, a(n-1)).$$
It is easy to see that (A_n) contains subsequences which converge to zero, whereas the sequence (A_n) itself does not converge to zero. By Corollary 6.11, the algebra $\mathcal{S}_{SO}(\mathbb{N})$ is not fractal. On the other hand, the *essential* fractality of these algebras is quite evident.

THEOREM 6.15. *The algebras $\mathcal{S}(\mathbb{N})$ and $\mathcal{S}(\mathbb{Z})$ are essentially fractal.*

PROOF. We prove the essential fractality of the algebra $\mathcal{S}(\mathbb{N})$ only. The proof for $\mathcal{S}(\mathbb{Z})$ proceeds similarly. Let $\mathbf{A} = (A_n)$ be a sequence in $\mathcal{S}(\mathbb{N})$ with $R_\eta(\mathbf{A}) = (A_{\eta(n)}) \in \mathcal{K}$ for some monotonically increasing sequence η. By Theorem 4.17, we can write the sequence $\mathbf{A}$ as $(P_n A P_n) + (K_n)$ with a band-dominated operator A and a sequence $(K_n) \in \mathcal{K}_0(\mathbb{N}) \subset \mathcal{S}(\mathbb{N}) \cap \mathcal{K}(\mathbb{N})$. Thus, letting n go to infinity in
$$A_{\eta(n)} = P_{\eta(n)} A P_{\eta(n)} + K_{\eta(n)}$$

yields the compactness of A via Proposition 4.2. Then $(P_n A P_n)$ is a compact sequence by Lemma 4.20, and hence the sequence $\mathbf{A}$ itself is compact by Theorem 4.18. By Theorem 6.6, the algebra $\mathcal{S}(\mathbb{N})$ is essentially fractal. □

CHAPTER 7

Applications

Here we discuss several applications of the notions and results introduced and proved above. Some of them are not restricted to the finite sections method for band-dominated operators and hold in a much wider context.

7.1. Approximation numbers

Let H be a separable Hilbert space. A family $(s_n)_{n \geq 1}$ of mappings $s_n : L(H) \to \mathbb{N}$ is called a *sequence of s-numbers* if, for arbitrary operators $A, B, C \in L(H)$,

- $\|A\| = s_1(A) \geq s_2(A) \geq \ldots \geq 0$.
- $s_n(A+B) \leq s_n(A) + \|B\|$.
- $s_n(ABC) \leq \|A\| \, s_n(B) \, \|C\|$.
- If $\operatorname{rank} A < n$, then $s_n(A) = 0$.
- $s_n(P) = 1$ for each orthogonal projection P of rank one.

One example of a sequence of s-numbers is provided by the approximation numbers

$$\Sigma_n(A) := \inf \{\|A - F\| : F \in L(H), \operatorname{rank} F < n\}$$

already mentioned in Section 4.2 (see [**28, 29**]). There is a related notion of an approximation number of a bounded sequence of matrices which is defined as follows. For $\mathbf{A} \in \mathcal{F}$ and for each $n \in \mathbb{N}$, the nth *approximation number* of $\mathbf{A}$ is the quantity

(7.1) $$\Sigma_n(\mathbf{A}) := \inf \{\|\mathbf{A} - \mathbf{F}\|_{\mathcal{F}} : \mathbf{F} \in \mathcal{K}, \operatorname{ess\,rank} \mathbf{F} < n\}.$$

Evidently,

(7.2) $$\Sigma_1(\mathbf{A}) = \|\mathbf{A} + \mathcal{G}\|_{\mathcal{F}/\mathcal{G}}$$

and, since the sequences of finite essential rank are dense in $\mathcal{K}$,

(7.3) $$\lim_{n \to \infty} \Sigma_n(\mathbf{A}) = \|\mathbf{A} + \mathcal{K}\|_{\mathcal{F}/\mathcal{K}}$$

for each sequence $\mathbf{A} \in \mathcal{F}$. The relation between the approximation numbers of a sequence (A_n) and the approximation numbers of the matrices A_n is as follows.

THEOREM 7.1. *For each sequence $\mathbf{A} = (A_n) \in \mathcal{F}$ and each positive integer k,*

(7.4) $$\limsup_{n \to \infty} \Sigma_k(A_n) = \Sigma_k(\mathbf{A}).$$

PROOF. Let $\mathbf{A} = (A_n) \in \mathcal{F}$ and $k \geq 1$, and let $\varepsilon > 0$ be arbitrarily given. For each $n \in \mathbb{N}$, choose an $n \times n$-matrix K_n with $\operatorname{rank} K_n \leq k - 1$ such that

(7.5) $$\|A_n - K_n\| \leq \Sigma_k(A_n) + \varepsilon.$$

The right-hand side of this estimate is bounded by $\|A_n\| + \varepsilon \leq \|\mathbf{A}\| + \varepsilon$, hence, the sequence $\mathbf{K} := (K_n)$ belongs to $\mathcal{F}$, and its essential rank is at most $k - 1$. The

estimate (7.5) yields further that
$$\limsup_{n\to\infty} \|A_n - K_n\| \leq \limsup_{n\to\infty} \Sigma_k(A_n) + \varepsilon,$$
whence via the limsup-formula (2.6) for the norm in $\mathcal{F}/\mathcal{G}$
$$\inf_{\mathbf{G}\in\mathcal{G}} \|\mathbf{A} - \mathbf{K} - \mathbf{G}\| \leq \limsup_{n\to\infty} \Sigma_k(A_n) + \varepsilon.$$
Thus, $\Sigma_k(\mathbf{A}) \leq \limsup_{n\to\infty} \Sigma_k(A_n) + \varepsilon$, and letting ε go to zero one finds
(7.6) $$\Sigma_k(\mathbf{A}) \leq \limsup_{n\to\infty} \Sigma_k(A_n).$$

For the reverse inequality, let again $\varepsilon > 0$ and choose sequences $\mathbf{G} \in \mathcal{G}$ and $\mathbf{K} = (K_n) \in \mathcal{F}$ with $\sup_n \operatorname{rank} K_n \leq k-1$ such that
$$\|\mathbf{A} - \mathbf{K} - \mathbf{G}\| \leq \Sigma_k(\mathbf{A}) + \varepsilon.$$
The left-hand side of this estimate is not less than $\|\mathbf{A} - \mathbf{K} + \mathcal{G}\|_{\mathcal{F}/\mathcal{G}}$ whence via the limsup-formula
$$\limsup_{n\to\infty} \|A_n - K_n\| \leq \Sigma_k(\mathbf{A}) + \varepsilon.$$
Since $\Sigma_k(A_n) \leq \|A_n - K_n\|$, we obtain
$$\limsup_{n\to\infty} \Sigma_k(A_n) \leq \Sigma_k(\mathbf{A}) + \varepsilon.$$
This holds for every $\varepsilon > 0$ which yields the estimate opposite to (7.6). □

For $k = 1$, one rediscovers from (7.4) the limsup-formula (2.6) (which has been used in the proof several times). Notice also that, for each sequence $(A_n) \in \mathcal{F}$,
$$\lim_{k\to\infty} \limsup_{n\to\infty} \Sigma_k(A_n) = \|(A_n) + \mathcal{K}\|_{\mathcal{F}/\mathcal{K}}$$
as we have already noticed in Section 5.1.

7.2. Rank-preserving discretizations

Böttcher, Chithra and Namboodiri [10] proved that for the finite sections of an arbitrary bounded linear operator $A \in L(l^2(\mathbb{N}))$
$$\lim_{n\to\infty} \Sigma_k(P_n A P_n) = \Sigma_k(A).$$
We will derive their result in a slightly more general context. Let H be an infinite dimensional separable Hilbert space, and let (P_n) be a sequence of orthogonal projections P_n of rank n on H which converges strongly to the identity operator I. If we identify $\mathbb{C}^n$ with $\operatorname{im} P_n$ via a fixed basis, then it makes sense to consider C^*-subalgebra $\mathcal{F}^C$ of $\mathcal{F}$ which consists of all sequences $\mathbf{A} = (A_n)$ for which the sequence $(A_n P_n)$ converges *-strongly on H. The strong limit of that sequence will be denoted by $W(\mathbf{A})$. Thus, W is a *-homomorphism from $\mathcal{F}^C$ into $L(H)$.

Let $\mathcal{L}$ be a linear (not necessarily closed) subspace of $L(H)$. A bounded linear mapping $D: \mathcal{L} \to \mathcal{F}^C$ is called a *discretization* on $\mathcal{L}$ if
$$W(D(A)) = A \quad \text{for each } A \in \mathcal{L}.$$
A discretization D on $\mathcal{L}$ is called *rank-preserving* if $\mathcal{L}$ contains all finite rank operators and if
$$\operatorname{ess\,rank} D(K) \leq \operatorname{rank} K \quad \text{for each } K \in \mathcal{L}.$$

For example, the finite sections discretization
$$D : L(H) \to \mathcal{F}^C, \quad A \mapsto (P_n A P_n)$$
is rank-preserving on $L(H)$.

PROPOSITION 7.2. *If $D : \mathcal{L} \to \mathcal{F}^C$ is a rank-preserving discretization, then*

(7.7) $$\operatorname{ess\,rank} D(K) = \operatorname{rank} K$$

for each operator K of finite rank.

This follows immediately from the estimate $\operatorname{rank} W(\mathbf{K}) \leq \operatorname{ess\,rank} \mathbf{K}$ holding for each sequence $\mathbf{K} \in \mathcal{F}^C$ due to Proposition 4.2 (a).

THEOREM 7.3. *Let $D : \mathcal{L} \to \mathcal{F}^C$ be a rank-preserving discretization with $\|D\| \leq 1$. Then*
$$\lim_{n \to \infty} \Sigma_k(D(A)_n) = \Sigma_k(A)$$
for each $A \in \mathcal{L}$ and each positive integer k.

The existence of the limit is part of the assertion.

PROOF. Given $\varepsilon > 0$ choose an operator $K \in L(H)$ with $\operatorname{rank} K \leq k - 1$ such that $\|A - K\| \leq \Sigma_k(A) + \varepsilon$. Since D is rank-preserving, one has $K \in \mathcal{L}$ and $\operatorname{ess\,rank} D(K) \leq k - 1$, and since D has norm not greater than one,
$$\|D(A)_n - D(K)_n\| \leq \|D(A - K)\|_{\mathcal{F}} \leq \|A - K\| \leq \Sigma_k(A) + \varepsilon.$$
Write $D(K)$ as $(F_n) + (G_n)$ with $\sup_n \operatorname{rank} F_n \leq k - 1$ and $(G_n) \in \mathcal{G}$ to obtain
$$\|D(A)_n - F_n - G_n\| \leq \Sigma_k(A) + \varepsilon,$$
whence
$$\Sigma_k(D(A)_n) \leq \|D(A)_n - F_n\| \leq \Sigma_k(A) + \varepsilon + \|G_n\|.$$
Consequently,
$$\limsup_{n \to \infty} \Sigma_k(D(A)_n) \leq \Sigma_k(A) + \varepsilon,$$
and since $\varepsilon > 0$ can be chosen arbitrarily,

(7.8) $$\limsup_{n \to \infty} \Sigma_k(D(A)_n) \leq \Sigma_k(A).$$

If $\Sigma_k(A) = 0$, then (7.8) implies $\lim_{n \to \infty} \Sigma_k(D(A)_n) = 0$ which is the assertion. So let $\Sigma_k(A) =: d > 0$. Contrary to what we want to show, assume that there is a $c \in (0, d)$ such that $\Sigma_k(D(A)_{n_j}) \leq c$ for infinitely many indices n_j. Then, for each $\varepsilon \in (0, d - c)$ and each index n_j, there is an operator $F_{n_j} \in L(\operatorname{im} P_{n_j}) \subseteq L(H)$ with $\operatorname{rank} F_{n_j} \leq k - 1$ and
$$\|D(A)_{n_j} - F_{n_j}\| < c + \varepsilon.$$
With Lemma 4.1, we get an operator $F \in L(H)$ with $\operatorname{rank} F \leq k - 1$ such that, for arbitrarily given vectors $x, y \in H$, the number $\langle Fx, y \rangle$ is a partial limit of the sequence $(\langle F_{n_j} x, y \rangle)_{j \geq 1}$. In particular, if x and y are chosen as unit vectors, then
$$|\langle D(A)_{n_j} x, y \rangle - \langle F_{n_j} x, y \rangle| \leq \|D(A)_{n_j} - F_{n_j}\| < c + \varepsilon.$$
Passing to a suitable subsequence and letting j go to infinity we obtain
$$|\langle Ax, y \rangle - \langle Fx, y \rangle| \leq c + \varepsilon.$$
This implies that $d \leq \|A - F\| \leq c + \varepsilon$ which contradicts the choice of ε. Hence,
$$\liminf_{n \to \infty} \Sigma_k(D(A)_n) \geq d = \Sigma_k(A).$$

Together with (7.8), this implies the assertion in case $d > 0$. □

COROLLARY 7.4. *Let $D : \mathcal{L} \to \mathcal{F}^C$ be a rank-preserving discretization with $\|D\| \leq 1$. Then*
$$\Sigma_k(D(A)) = \Sigma_k(A)$$
for each $A \in \mathcal{L}$ and each positive integer k.

COROLLARY 7.5. *Let $D : \mathcal{L} \to \mathcal{F}^C$ be a rank-preserving discretization with $\|D\| \leq 1$. Then*
$$\Sigma_k(D(A)) = \inf\{\|D(A) - D(F)\|_{\mathcal{F}} : F \in \mathcal{L} \text{ with } \operatorname{rank} F \leq k - 1\}$$
for each $A \in \mathcal{L}$ and each positive integer k.

In contrast to the definition of the approximation number of a sequence, the infimum is taken here only over sequences of rank less than k which arise from discretization by D. This result suggests that approximation numbers (of operators on a Hilbert space and of approximation sequences in $\mathcal{F}$ as well) might possess an inverse closedness property. I must leave this question open.

PROOF OF COROLLARY 7.5. Let $A \in \mathcal{L}$, $\varepsilon > 0$, $k \geq 1$, and choose $K \in \mathcal{L}$ with $\operatorname{rank} K \leq k - 1$ such that
$$\|A - K\| \leq \Sigma_k(A) + \varepsilon.$$
Then the sequence $D(K)$ is of essential rank less than k, and since $\|D\| \leq 1$ one has
$$\inf\{\|D(A) - D(F)\|_{\mathcal{F}} : F \in \mathcal{L} \text{ with } \operatorname{rank} F \leq k - 1\} \leq \|A - K\| \leq \Sigma_k(A) + \varepsilon.$$
Letting ε go to zero and taking into account that $\Sigma_k(A) = \Sigma_k(D(A))$ by Corollary 7.4 one gets the assertion. □

Notice that this theorem and its corollaries remain valid if we do not require that all finite rank operators belong to $\mathcal{L}$ but merely suppose that each operator of rank k can be approximated as closely as desired by operators in $\mathcal{L}$ of rank at most k. This observation makes these results applicable, for example, to collocation and quadrature methods for integral equations on L^2-spaces (where point evaluation makes sense only on a dense subset of L^2; see [**22, 30, 31**] and the references therein).

7.3. Arveson dichotomy: band-dominated operators

Given a self-adjoint matrix A and an open interval $U \subseteq \mathbb{R}$, we let $N(\sigma(A) \cap U)$ denote the number of the eigenvalues of A which lie in U, counted with respect to their multiplicity. Let $(A_n) \in \mathcal{F}$ be a self-adjoint sequence. A point $\lambda \in \mathbb{R}$ is called *essential* for the sequence (A_n) if, for every open interval U containing λ,
$$\lim_{n \to \infty} N(\sigma(A_n) \cap U) = \infty,$$
and $\lambda \in \mathbb{R}$ is *transient* for the sequence (A_n) if there is an open interval U containing λ such that
$$\sup_n N(\sigma(A_n) \cap U) < \infty.$$
Further one says that a self-adjoint sequence $(A_n) \in \mathcal{F}$ enjoys the *Arveson dichotomy* if every real number is either essential or transient for this sequence. Arveson introduced and studied these notions in a series of papers [**2, 3, 4**]. He proved

the Arveson dichotomy of the sequence $(P_n A P_n)$ in case A is a self-adjoint band operator, and he extended this result to a class of self-adjoint band-dominated operators which satisfy both a Wiener and a Besov space condition (see also Section 7.1 in [**23**]). The goal of this section is to prove the Arveson dichotomy for general self-adjoint band-dominated operators. Our proof will rest on a straightforward application of Theorem 5.1 (c) characterizing Fredholm sequences in terms of singular values and of Theorem 5.8 stating the Fredholmness of the finite sections sequence $(P_n A P_n)$ for every Fredholm band-dominated operator A.

THEOREM 7.6. *Let A be a self-adjoint band-dominated operator on $l^2(\mathbb{N})$. Then the sequence $(P_n A P_n)$ has the Arveson dichotomy, and the set of all essential points of that sequence coincides with the essential spectrum of A.*

One part of the proof is based upon the following proposition which holds in a more general context. In particular, this result establishes the existence of essential points. Again, $\mathcal{F}^C$ stands for the algebra of all *-strongly convergent sequences.

PROPOSITION 7.7. *Let $(A_n) \in \mathcal{F}^C$ be a sequence of self-adjoint matrices with strong limit A. Then every point in the essential spectrum of A is an essential point for the sequence (A_n).*

PROOF. Let $\lambda \in \mathbb{R}$ be not essential for (A_n). We have to show that then $A - \lambda I$ is a Fredholm operator.

By assumption, there are an infinite set $\mathbb{M}$ of positive integers as well as an interval $U = (\lambda - \varepsilon, \lambda + \varepsilon)$ with $\varepsilon > 0$ such that
$$\sup_{n \in \mathbb{M}} N(\sigma(A_n) \cap U) =: k < \infty.$$

For $n \in \mathbb{M}$, let T_n denote the orthogonal projection from $l^2(\mathbb{N})$ onto the U-spectral subspace of A_n (where we again identify $\mathbb{C}^n$ with the subspace im P_n of $l^2(\mathbb{N})$). The rank of the projection T_n is not greater than k. It is moreover obvious that the operators
$$B_n := (A_n - \lambda P_n)(I - T_n) + T_n$$
are invertible for all $n \in \mathbb{M}$ and that their inverses are uniformly bounded by $\max \{1/\varepsilon, 1\}$. Hence,

(7.9) $$(A_n - \lambda P_n)(I - T_n)B_n^{-1} = I - T_n B_n^{-1}$$

for all $n \in \mathbb{M}$. Now we argue as in the proof of Corollary 5.11. Employing the sequential compactness of the unit ball in $L(l^2(\mathbb{N}))$ with respect to weak convergence, we choose weakly convergent subsequences $((I - T_{n_r})B_{n_r}^{-1})_{r \geq 1}$ of $((I - T_n)B_n^{-1})_{n \in \mathbb{M}}$ and $(T_{n_r} B_{n_r}^{-1})_{r \geq 1}$ of $(T_n B_n^{-1})_{n \in \mathbb{M}}$ with limits B and T, respectively. The product of a weakly convergent sequence with limit C and a *-strongly convergent sequence with limit D is weakly convergent with limit CD. Thus, passing to subsequences and taking the weak limit in (7.9) yields $(A - \lambda I)B = I - T$. Furthermore, the rank of T is not greater than k by Proposition 4.2 (a). Thus, $(A - \lambda I)B - I$ is a compact operator. Analogously, the compactness of $B(A - \lambda I) - I$ follows, whence the Fredholmness of A. □

Arveson constructed an example which illustrates that the inclusion in Proposition 7.7 can be proper. Specifically, he constructed a self-adjoint unitary operator $A \in L(l^2(\mathbb{N}))$ with

(7.10) $$\sigma(A) = \sigma_{ess}(A) = \{-1, 1\}$$

such that 0 is an essential point of the sequence $(P_n A P_n)$. Here is his example for completeness.

EXAMPLE. Let $E_1 := 4\mathbb{N}$ and $E_2 := 2\mathbb{N} \setminus 4\mathbb{N}$, and define a function f on E_1 by
$$f(k) = k^2 + 1 \quad (\in 2\mathbb{N} - 1).$$
Let $O_1 := f(E_1)$, and set $O_2 := (2\mathbb{N} - 1) \setminus O_1$. Clearly, both E_2 and O_2 are infinite subsets of $\mathbb{N}$, and we define f on E_2 to be any bijection from E_2 onto O_2. This construction implies a permutation π of $\mathbb{N}$ with $\pi^2 = \mathrm{Id}$ via
$$\pi(k) := \begin{cases} f(k) & \text{if } k \text{ even} \\ f^{-1}(k) & \text{if } k \text{ odd.} \end{cases}$$
We claim that the operator
$$A : l^2(\mathbb{N}) \to l^2(\mathbb{N}), \quad (a_n)_{n \in \mathbb{N}} \mapsto (a_{\pi(n)})_{n \in \mathbb{N}}$$
has the properties announced.

Evidently, $A = A^* = A^{-1} \neq \pm I$, whence (7.10) easily follows. In order to see that 0 is an essential point of $(P_n A P_n)$, let $\mathbb{N}_n$ denote the set $\{1, 2, \ldots, n\}$ and write $\sharp S$ for the number of the elements of a set S. It is elementary to check that $\sharp(f(E_1 \cap \mathbb{N}_n) \setminus \mathbb{N}_n)$ tends to infinity as $n \to \infty$, which implies that
$$\lim_{n \to \infty} \sharp(f(E \cap \mathbb{N}_n) \setminus \mathbb{N}_n) = \infty$$
and, consequently,

(7.11) $$\lim_{n \to \infty} \sharp(\pi(\mathbb{N}_n) \setminus \mathbb{N}_n) = \infty.$$

Since A maps the basis element $e_k := (0, \ldots, 0, 1, 0, \ldots)$ of $l^2(\mathbb{N})$ (with the 1 standing at the kth place) to $e_{\pi(k)}$, it follows that $P_n A P_n e_k = 0$ for every k belonging to the set $S_n := \{k \in \mathbb{N}_n : \pi(k) \notin \mathbb{N}_n\}$. But the cardinality of S_n tends to infinity as $n \to \infty$ due to (7.11). Hence, 0 is an eigenvalue of $P_n A P_n$ for all sufficiently large n, and the multiplicity of this eigenvalue tends to infinity. Consequently, 0 is an essential point of $(P_n A P_n)$. □

PROOF OF THEOREM 7.6. We claim that every point in $\mathbb{R} \setminus \sigma_{ess}(A)$ is transient, which implies both assertions of Theorem 7.6. Indeed, every point in $\sigma_{ess}(A)$ is essential by Proposition 7.7. If there were an essential point of $(P_n A P_n)$ which is not in $\sigma_{ess}(A)$ then the claim would give that this point is transient, which is impossible. Hence, the set of all essential points of $(P_n A P_n)$ coincides with the essential spectrum of A. The other assertion of Theorem 7.6 follows easily since all points in $\mathbb{R} \setminus \sigma_{ess}(A)$ are transient by our claim again.

To prove the claim, we have to show that if $\lambda \in \mathbb{R} \setminus \sigma_{ess}(A)$, then λ is transient. Equivalently, we have to show that if $B := A - \lambda I$ is Fredholm, then 0 is a transient point for the sequence $(P_n B P_n)$. Let B be a Fredholm band-dominated operator. Then, by Theorem 5.8 (a), the sequence $(P_n B P_n)$ is Fredholm. Let k denote the α-number of that sequence. By Theorem 5.1 (c) and the definition of the α-number,
$$\liminf_{n \to \infty} \sigma_{k+1}(P_n B P_n) =: C > 0 \quad \text{and} \quad \liminf_{n \to \infty} \sigma_k(P_n B P_n) = 0.$$
Choose $U := (-C/2, C/2)$. Since the singular values of a self-adjoint matrix are just the absolute values of the eigenvalues of that matrix, we conclude that $N(\sigma(P_n B P_n) \cap U) \leq k$ for all n. Thus, 0 is transient. □

An obvious modification in the proof yields a slight generalization of Theorem 7.6.

THEOREM 7.8. *Let* $\mathbf{A} = (A_n)$ *be a self-adjoint sequence in* $\mathcal{S}(\mathbb{N})$ *with strong limit* A. *The sequence* (A_n) *has the Arveson dichotomy, and the set of all essential points of that sequence coincides with the essential spectrum of* A.

Basically, one has to show that if $B := A - \lambda I$ is Fredholm, then 0 is a transient point for the sequence $(B_n) := (A_n) - \lambda(I_n)$. This can be done as in the proof of Theorem 7.6 by referring to Theorem 5.8 (b) now. □

7.4. Arveson dichotomy: the general setting

Arveson dichotomy settles a relation between spectral quantities of a sequence and those of its subsequences. Indeed, a point $\lambda \in \mathbb{R}$ is *not* transient for a self-adjoint sequence (A_n) if and only if is is essential for a subsequence of (A_n). So one might expect that, in presence of essential fractality, every self-adjoint sequence shows Arveson dichotomy. We will see now that this is indeed true.

PROPOSITION 7.9. *Let* $\mathbf{A} = (A_n) \in \mathcal{F}$ *be a sequence of self-adjoint matrices. If* $\lambda \in \mathbb{R} \setminus \sigma(\mathbf{A} + \mathcal{K})$, *then* λ *is a transient point for the sequence* $\mathbf{A}$.

PROOF. If we denote $A_n - \lambda I_n$ by B_n, then we have to show that 0 is a transient point for the sequence (B_n). Since $\lambda \in \mathbb{R} \setminus \sigma(\mathbf{A} + \mathcal{K})$, the sequence (B_n) is Fredholm. Let k denote its α-number. By Theorem 5.1 (c) and the definition of the α-number,

$$\liminf_{n \to \infty} \sigma_{k+1}(B_n) =: C > 0 \quad \text{and} \quad \liminf_{n \to \infty} \sigma_k(B_n) = 0.$$

Choose $U := (-C/2, C/2)$. Since the singular values of a self-adjoint matrix are just the absolute values of the eigenvalues of that matrix, we conclude that $N(\sigma(B_n) \cap U) \leq k$ for all n. Thus, 0 is transient. □

PROPOSITION 7.10. *Let* $\mathbf{A} = (A_n) \in \mathcal{F}$ *be a sequence of self-adjoint matrices. If* $\lambda \in \sigma(\mathbf{A} + \mathcal{K})$, *then* λ *is not a transient point for the sequence* $\mathbf{A}$. *If, moreover, the sequence* $\mathbf{A}$ *belongs to an essentially fractal C^*-subalgebra of $\mathcal{F}$ which contains the identity element, then* λ *is an essential point for* $\mathbf{A}$.

PROOF. Let the point $\lambda \in \mathbb{R}$ be transient for (A_n). For the first assertion we have to show that $(A_n) - \lambda(P_n)$ is a Fredholm sequence. By assumption, there is an interval $U = (\lambda - \varepsilon, \lambda + \varepsilon)$ with $\varepsilon > 0$ such that

(7.12) $$\sup_{n \in \mathbb{N}} N(\sigma(A_n) \cap U) =: k < \infty.$$

Let T_n denote the orthogonal projection from $\mathbb{C}^n$ onto the U-spectral subspace of A_n. The rank of the projection T_n is not greater than k. It is moreover obvious that the operators

$$B_n := (A_n - \lambda P_n)(I - T_n) + T_n$$

are invertible for all $n \in \mathbb{N}$ and that their inverses are uniformly bounded by $\max\{1/\varepsilon, 1\}$. Hence,

(7.13) $$(A_n - \lambda P_n)(I - T_n)B_n^{-1} = I - T_n B_n^{-1}.$$

Since (T_n) is a compact sequence (of essential rank not greater than k), the identity (7.13) shows the invertibility from the right-hand side of the coset $(A_n - \lambda P_n) + \mathcal{K}$ in $\mathcal{F}/\mathcal{K}$. Since this coset is self-adjoint, it is invertible, i.e., $(A_n - \lambda P_n)$ is a Fredholm sequence. This settles the first assertion. For the second one assume that $\lambda \in \mathbb{R}$ is not essential for the sequence (A_n). Then (7.13) holds only for infinitely many of the n; thus, there is a subsequence of $(A_n - \lambda P_n)$ which is Fredholm. By Corollary 6.12, the sequence $(A_n - \lambda P_n)$ itself is Fredholm. □

THEOREM 7.11. *Let* $\mathbf{A} = (A_n) \in \mathcal{F}$ *be a sequence of self-adjoint matrices. Then*

(a) *the spectrum of the coset* $\mathbf{A} + \mathcal{K}$ *in the algebra* $\mathcal{F}/\mathcal{K}$ *consists exactly of those real numbers which are not transient for* $\mathbf{A}$.

(b) *if, moreover, the sequence* $\mathbf{A}$ *belongs to an essentially fractal* C^**-subalgebra of* $\mathcal{F}$ *which contains the identity element, then this sequence has Arveson dichotomy, i.e., the set of the non-transient points for* $\mathbf{A}$ *coincides with the set of the essential points for* $\mathbf{A}$.

This follows easily from the previous propositions. In [**23**], Theorem 7.12, the Arveson dichotomy is settled for a class of approximation sequences which belong a so-called standard subalgebra of $\mathcal{F}$. Without going into detail here, let me only mention that it is not hard to check that every standard algebra (in the sense of Section 6.1.1 in [**23**]) is indeed essentially fractal. Thus, Theorem 7.12 from [**23**] becomes a special case of the above theorem.

7.5. Essential spectra

The results of the previous sections imply that the essential spectrum of a self-adjoint band-dominated operator A can be determined by considering convergent sequences of eigenvalues of $P_n A P_n$ with increasing multiplicity. For non-self-adjoint sequences this is definitely impossible as the finite sections of the (forward or backward) shift operator show. Indeed, in this case 0 is the only eigenvalue of $P_n A P_n$, whereas the essential spectrum of A is the unit circle. Nevertheless, one can determine the essential spectrum of an (in general, non-self-adjoint) band-dominated operator A from eigenvalues with large multiplicity if one considers perturbations of the finite sections sequence $(P_n A P_n)$. In fact, perturbations by zero sequences will suffice.

THEOREM 7.12. *Let* $(A_n) \in \mathcal{F}$. *Then the following assertions are equivalent for* $\lambda \in \mathbb{C}$:

(a) $\lambda \in \sigma((A_n) + \mathcal{K})$.

(b) *There are a sequence* $(G_n) \in \mathcal{G}$ *and a monotonically increasing sequence* $n_1 < n_2 < \ldots$ *of positive integers such that, for each* $k \in \mathbb{N}$, *there are eigenvalues* $\lambda_k \in \sigma(A_{n_k} - G_{n_k})$ *with*
$$\lambda_k \to \lambda \quad \text{and} \quad \lim_{k \to \infty} N(\lambda_k) = \infty.$$

Here, $N(\lambda_k)$ stands for the geometric multiplicity of the eigenvalue λ_k of $A_{n_k} - G_{n_k}$, i.e., for the kernel dimension of $A_{n_k} - G_{n_k} - \lambda_k I_{\delta(n_k)}$ where δ refers to the dimension function of $\mathcal{F}$.

PROOF. It is sufficient to prove the assertion in case $\lambda = 0$. We start with the implication (a) $\Rightarrow$ (b). Assume that $(A_n) \in \mathcal{F}$ is not a Fredholm sequence, and let
$$A_n = E_n^* \text{diag}\,(\sigma_1(A_n), \ldots, \sigma_{\delta(n)}(A_n)) F_n$$
refer to the singular value decomposition of A_n. Since (A_n) is not Fredholm,
$$\liminf_{n \to \infty} \sigma_k(A_n) = 0 \quad \text{for each } k \in \mathbb{N}$$

by Theorem 5.1. Hence there is a strongly monotonically increasing sequence $(n_k)_{k \geq 1}$ of positive integers such that
$$\sigma_k(A_{n_k}) < 1/k \quad \text{for each } k \geq 1.$$
Let
$$G_{n_k} := E_{n_k}^* \text{diag}\,(\sigma_1(A_{n_k}), \ldots, \sigma_k(A_{n_k}), 0, \ldots, 0) F_{n_k}$$
with $n_k - k$ zeros in the right-hand part of the diagonal matrix, and set $G_n = 0$ for n being not of the form n_k for some k. Then $\|G_{n_k}\| < 1/k$. Hence, $(G_n) \in \mathcal{G}$, and 0 is a singular value of $A_{n_k} - G_{n_k}$ of multiplicity at least k. Since the kernels of A^*A and of A coincide for every bounded linear operator A on a Hilbert space, the point 0 is also an eigenvalue of $A_{n_k} - G_{n_k}$ of geometric multiplicity not less than k. Thus, one can choose $\lambda_k := 0$. Then $\lambda_k \to 0$ and $N(\lambda_k) \geq k$, which implies (b).

For the reverse implication we prove that if $0 \notin \sigma((A_n) + \mathcal{K})$, then there is no sequence $(G_n) \in \mathcal{G}$ for which there is a sequence of eigenvalues λ_k of $A_{n_k} - G_{n_k}$ with $|\lambda_k| \to 0$ and $N(\lambda_k) \to \infty$.

Since $(A_n) + \mathcal{K}$ is invertible, the sequence $(A_n) + (G_n)$ is Fredholm for each sequence $(G_n) \in \mathcal{G}$. Let $(A_{n_k} - G_{n_k})$ be a subsequence of $(A_n - G_n)$, and let $\lambda_k \in \sigma(A_{n_k} - G_{n_k})$ be eigenvalues with $|\lambda_k| \to 0$ as $k \to \infty$. Since the set of all Fredholm sequences is open, we can assume that $(A_{n_k} - G_{n_k} - \lambda_k I_{\delta(n_k)})$ is a Fredholm sequence, too (otherwise we consider a suitable of its subsequences for k being sufficiently large). Let r denote the α-number of that sequence. Then there are bounded sequences (C_{n_k}) and (J_{n_k}) of matrices with rank $J_{n_k} \leq r$ such that

(7.14) $$C_{n_k}(A_{n_k} - G_{n_k} - \lambda_k I_{\delta(n_k)}) = I_{\delta(n_k)} - J_{n_k}$$

for all $k \geq 1$. We estimate the geometric multiplicity of the eigenvalue λ_k of $A_{n_k} - G_{n_k}$. Let x belong to the kernel of $A_{n_k} - G_{n_k} - \lambda_k I_{\delta(n_k)}$. Then (7.14) implies that $(I_{\delta(n_k)} - J_{n_k})x = 0$, whence $x = J_{n_k}x$ and
$$\ker(A_{n_k} - G_{n_k} - \lambda_k I_{\delta(n_k)}) \subseteq \operatorname{im} J_{n_k}.$$
Thus,
$$N(\lambda_k) = \dim \ker(A_{n_k} - G_{n_k} - \lambda_k I_{\delta(n_k)}) \leq \operatorname{rank} J_{n_k} \leq r,$$
i.e., the geometric multiplicities of the λ_k are bounded by r uniformly with respect to k. □

There is a close relative of the preceding theorem which says that, if multiplicities are *not* taken into account, then one obtains the spectrum of (A_n) modulo $\mathcal{G}$ instead of the spectrum modulo $\mathcal{K}$ (i.e., the stability spectrum of (A_n) instead of the essential spectrum). For details see Theorem 3.19 in [**23**].

In case A is a band-dominated operator then
$$\sigma(A + K(l^2(\mathbb{N}))) = \sigma((P_n A P_n) + \mathcal{K})$$
by Theorem 5.8. Thus, Theorem 7.12 offers the possibility to determine the essential spectrum of a band-dominated operator asymptotically.

COROLLARY 7.13. *Let $A \in \mathcal{A}(\mathbb{N})$. Then the following assertions are equivalent for $\lambda \in \mathbb{C}$:*
(a) *The operator $A - \lambda I$ is not Fredholm.*
(b) *There are a sequence $(G_n) \in \mathcal{G}$ and a monotonically increasing sequence $n_1 <$*

$n_2 < \ldots$ of positive integers such that, for each $k \in \mathbb{N}$, there are eigenvalues $\lambda_k \in \sigma(P_{n_k} A P_{n_k} - G_{n_k})$ with

$$\lambda_k \to \lambda \quad \text{and} \quad \lim_{k \to \infty} N(\lambda_k) = \infty.$$

An analogous result holds for band-dominated operators on $l^2(\mathbb{Z})$ and the corresponding finite sections.

7.6. Essential pseudospectra

Let $\varepsilon > 0$. An element a of a C^*-algebra $\mathcal{B}$ with identity element e is said to be ε-*invertible* if it is invertible and $\|a^{-1}\| < 1/\varepsilon$. The ε-*pseudospectrum* $\sigma^{(\varepsilon)}(a)$ of a consists of all $\lambda \in \mathbb{C}$ for which $a - \lambda e$ is not ε-invertible. An equivalent description of pseudospectra reads as follows.

THEOREM 7.14. *For every $a \in \mathcal{B}$ and $\varepsilon > 0$,*

$$\sigma^{(\varepsilon)}(a) = \cup_{r \in \mathcal{B}: \|r\| \leq \varepsilon} \sigma(a - p).$$

A proof can be found in [**46**], [**23**] (Theorem 3.27) and [**13**], for instance. See also the recent textbook [**57**] for an exhaustive account on pseudospectra.

Pseudospectra are distinguished by their excellent convergence properties. The standard example discussed in [**9, 13, 15, 23**] is provided by the finite sections method for Toeplitz operators. Let $a(t) = t$ for $t \in \mathbb{T}$. Then the only eigenvalue of the Toeplitz matrix $P_n T(a) P_n$ is $0 \in \mathbb{C}$ which is fairly different from the spectrum of the Toeplitz operator $T(a)$ which is the closed unit disc. On the other hand, the ε-pseudospectra of the matrices $P_n T(a) P_n$ converge with respect to the Hausdorff metric to the ε-pseudospectrum of $T(a)$ for each $\varepsilon > 0$ as has been first proved in [**9**]. The convergence of the ε-pseudospectra of the matrices $P_n A P_n$ for an arbitrary band-dominated operator A on $l^2(\mathbb{N})$ has been considered in [**33**] and [**37**], Section 6.3.

Here we will discuss the approximation of essential pseudospectra. We refer to the ε-invertibility and the ε-pseudospectrum of the coset $(A_n) + \mathcal{K}$ in $\mathcal{F}/\mathcal{K}$ as the *essential ε-invertibility* and the *essential ε-pseudospectrum* of the sequence $(A_n) \in \mathcal{F}$. As in the previous subsections (which are concerned with essential spectra of self-adjoint sequences in $\mathcal{F}$), the points in $\sigma^{(\varepsilon)}((A_n) + \mathcal{K})$ can be related with pseudoeigenvalues of A_n of high multiplicity.

THEOREM 7.15. *Let $(A_n) \in \mathcal{F}$ and $\varepsilon > 0$. The following assertions are equivalent for $\lambda \in \mathbb{C}$:*

(a) $\lambda \in \sigma^{(\varepsilon)}((A_n) + \mathcal{K})$.

(b) *There are a sequence $(R_n) \in \mathcal{F}$ with $\|(R_n)\| \leq \varepsilon$ and a monotonically increasing sequence $n_1 < n_2 < \ldots$ of positive integers such that, for each $k \in \mathbb{N}$, there are eigenvalues $\lambda_k \in \sigma(A_{n_k} - R_{n_k})$ with*

$$\lambda_k \to \lambda \quad \text{and} \quad \lim_{k \to \infty} N(\lambda_k) = \infty.$$

Again, $N(\lambda_k)$ stands for the geometric multiplicity of the eigenvalue λ_k of $A_{n_k} - R_{n_k}$.

A part of the proof of Theorem 7.15 is based on the following result which had been conjectured by Böttcher and was proved by Daniluk [**9, 15, 23**].

THEOREM 7.16. **(Daniluk)** *Let $\mathcal{B}$ be a C^*-algebra with identity e, let $a \in \mathcal{B}$, and suppose $a - \lambda e$ to be invertible for all λ in some open subset U of the complex plane. If $\|(a - \lambda e)^{-1}\| \leq C$ for all $\lambda \in U$, then $\|(a - \lambda e)^{-1}\| < C$ for all $\lambda \in U$.*

Otherwise stated: the analytic function $U \to \mathcal{B}$, $\lambda \mapsto (a - \lambda e)^{-1}$ is subject to the maximum principle. This is quite surprising since – in contrast to complex-valued analytic functions – the maximum principle does not hold for general operator-valued analytic functions as simple examples show. The consequence of Daniluk's theorem which is used in the following proof is that each open neighborhood of an ε-pseudoeigenvalue $\lambda \in \sigma^{(\varepsilon)}(a)$ contains a $\lambda_0 \in \sigma^{(\varepsilon)}(a)$ with
$$\|(a - \lambda_0 e)^{-1}\| > 1/\varepsilon.$$
Indeed, otherwise one would have $\|(a - \lambda_0 e)^{-1}\| \leq 1/\varepsilon$ for all λ_0 in a certain open neighborhood U of λ. By Theorem 7.16, this implies $\|(a - \lambda_0 e)^{-1}\| < 1/\varepsilon$ for all $\lambda_0 \in U$ which is impossible since $\lambda \in U$. In particular, ε-pseudospectra cannot contain isolated points.

PROOF OF THEOREM 7.15. Parts of this proof are close to the proof of Theorem 7.12. Again, it is sufficient to prove the assertion in case $\lambda = 0$. We start with the implication $(a) \Rightarrow (b)$. Thus, we let $0 \in \sigma^{(\varepsilon)}((A_n) + \mathcal{K})$. First consider the case when $(A_n) \in \mathcal{F}$ is not a Fredholm sequence. Let
$$A_n = E_n^* \mathrm{diag}\,(\sigma_1(A_n), \ldots, \sigma_{\delta(n)}(A_n)) F_n$$
stand for the singular value decomposition of A_n. Since (A_n) is not Fredholm,
$$\liminf_{n \to \infty} \sigma_k(A_n) = 0 \quad \text{for each } k \in \mathbb{N}.$$
Choose a strongly monotonically increasing sequence $(n_k)_{k \geq 1}$ with
$$\sigma_k(A_{n_k}) < \varepsilon \quad \text{for each } k \geq 1$$
and set
$$R_{n_k} := E_{n_k}^* \mathrm{diag}\,(\sigma_1(A_{n_k}), \ldots, \sigma_k(A_{n_k}), 0, \ldots, 0) F_{n_k}$$
with $n_k - k$ zeros in the right-hand part of the diagonal matrix. Then $\|R_{n_k}\| < \varepsilon$, and 0 is a singular value of $A_{n_k} - R_{n_k}$ of multiplicity at least k. Then 0 is also an eigenvalue of $A_{n_k} - R_{n_k}$ of geometric multiplicity not less than k. Thus, if we choose $\lambda_k := 0$ then we get $N(\lambda_k) \geq k$ which implies (b).

Now assume that (A_n) is a Fredholm sequence but that

(7.15) $$\|((A_n) + \mathcal{K})^{-1}\| \geq 1/\varepsilon.$$

By Daniluk's theorem, each open neighborhood of 0 contains a $\mu_0 \in \mathbb{C}$ such that
$$\|((A_n) - \mu_0(I_{\delta(n)}) + \mathcal{K})^{-1}\| > 1/\varepsilon.$$
In particular, for each $r \geq 1$, there is a $\mu_r \in \mathbb{C}$ such that $\mu_r \to 0$ as $r \to \infty$ and
$$\|((A_n) - \mu_r(I_{\delta(n)}) + \mathcal{K})^{-1}\| \geq \left(\varepsilon - \frac{1}{r}\right)^{-1}.$$
By Corollary 5.3, this implies
$$\sup_k \liminf_{n \to \infty} \sigma_k(A_n - \mu_r I_{\delta(n)}) \leq \varepsilon - \frac{1}{r}.$$

for every $r \geq 1$. Thus,

$$\liminf_{n \to \infty} \sigma_k(A_n - \mu_r I_{\delta(n)}) \leq \varepsilon - \frac{1}{r} \tag{7.16}$$

for every k, $r \geq 1$. For each $k \geq 1$, choose $r_k \geq 1$ such that $|\mu_{r_k}| < 1/k$. Then, by (7.16),

$$\liminf_{n \to \infty} \sigma_k(A_n - \mu_{r_k} I_{\delta(n)}) \leq \varepsilon - \frac{1}{r_k}$$

for each $k \geq 1$. Further choose a sequence $n_1 < n_2 < \ldots$ such that

$$\sigma_k(A_{n_k} - \mu_{r_k} I_{\delta(n_k)}) \leq \varepsilon - \frac{1}{2r_k} \tag{7.17}$$

for each $k \geq 1$. Let $E_{n_k}^* \operatorname{diag}(\sigma_1, \ldots, \sigma_{\delta(n_k)}) F_{n_k}$ refer to the singular value decomposition of $A_{n_k} - \mu_{r_k} I_{\delta(n_k)}$ and set

$$R_{n_k} := E_{n_k}^* \operatorname{diag}(\sigma_1, \ldots, \sigma_k, 0, \ldots, 0) F_{n_k}.$$

Then, by (7.17),

$$\|R_{n_k}\| \leq \varepsilon - \frac{1}{2r_k} < \varepsilon$$

for all $k \geq 1$, and 0 is a k-fold singular value of $A_{n_k} - R_{n_k} - \mu_{r_k} I_{\delta(n_k)}$. Thus, μ_{r_k} is a k-fold singular value of $A_{n_k} - R_{n_k}$ and, hence, an eigenvalue of $A_{n_k} - R_{n_k}$ of geometric multiplicity at least k. So we can choose $\lambda_k := \mu_{r_k}$ to get $|\lambda_k| < 1/k \to 0$ and $N(\lambda_k) \geq k \to \infty$.

To prove the reverse implication $(b) \Rightarrow (a)$ we verify that if $0 \notin \sigma^{(\varepsilon)}((A_n) + \mathcal{K})$, then there is *no* sequence (R_n) with $\|(R_n)\| \leq \varepsilon$ for which there is a sequence of eigenvalues λ_k of $A_{n_k} - R_{n_k}$ with $|\lambda_k| \to 0$ and $N(\lambda_k) \to \infty$.

Since $(A_n) + \mathcal{K}$ is ε-invertible, there is a $\delta > 0$ such that

$$\|((A_n) + \mathcal{K})^{-1}\| = \frac{1}{\varepsilon + \delta}.$$

Let $(R_n) \in \mathcal{F}$ be a sequence with $\|(R_n)\| \leq \varepsilon$. A standard Neumann series argument implies that $(A_n) - (R_n)$ is a Fredholm sequence with

$$\|((A_n) - (R_n) + \mathcal{K})^{-1}\| \leq \frac{1}{\delta}.$$

Let $(A_{n_k} - R_{n_k})$ be a subsequence of $(A_n - R_n)$, and let $\lambda_k \in \sigma(A_{n_k} - R_{n_k})$ be eigenvalues with $|\lambda_k| \to 0$ as $k \to \infty$. Clearly, $(A_{n_k} - R_{n_k})$ is a Fredholm sequence, too (in an appropriately chosen algebra $\mathcal{F}$ of matrix sequences). Since the set of all Fredholm sequences is open, we can further assume that $(A_{n_k} - R_{n_k} - \lambda_k I_{\delta(n_k)})$ is a Fredholm sequence (otherwise we consider a suitable of its subsequences for k being sufficiently large). As in the proof of Theorem 7.12, one can show that the geometric multiplicities of the λ_k are uniformly bounded. □

EXAMPLE. This example shows that Theorem 7.15 becomes wrong if one replaces in its formulation geometric multiplicity by algebraic multiplicity, i.e., by the multiplicity of λ_k as a zero of the characteristic polynomial of $A_{n_k} - R_{n_k}$. Consider the $n \times n$-matrices

$$A_n := \begin{pmatrix} 0 & 1 & & & \\ & 0 & 1 & & \\ & & \ddots & \ddots & \\ & & & 0 & 1 \\ & & & & 0 \end{pmatrix} \quad \text{and} \quad K_n := \begin{pmatrix} 0 & 0 & \cdots & 0 \\ \vdots & \vdots & & \vdots \\ 0 & 0 & \cdots & 0 \\ 1 & 0 & \cdots & 0 \end{pmatrix}.$$

Then (A_n) is a Fredholm sequence with α-number 1, (K_n) is a compact sequence of essential rank 1, and $(A_n) + (K_n)$ is a sequence of isometries. Thus,

$$\|(A_n)+\mathcal{K}\| \leq \|(A_n)+(K_n)\| = 1 \quad \text{and} \quad \|((A_n)+\mathcal{K})^{-1}\| \leq \|((A_n)+(K_n))^{-1}\| = 1,$$

whence

$$\|(A_n) + \mathcal{K}\| = \|((A_n) + \mathcal{K})^{-1}\| = 1.$$

(This equality follows also easily from Theorem 5.8 (b).) In particular, the coset $(A_n) + \mathcal{K}$ is ε-invertible for each $\varepsilon < 1$. If we choose $(R_n) := 0$, then 0 becomes an eigenvalue of $A_n - R_n = A_n$ of algebraic multiplicity n (whereas its geometric multiplicity is 1). □

In the setting of essentially fractal subalgebras of $\mathcal{F}$ (and in particular for the finite sections algebra of the band-dominated operators), Theorem 7.15 can be completed as follows.

PROPOSITION 7.17. (a) Let $\mathcal{B}$ be an essentially fractal subalgebra of $\mathcal{F}$ which contains the identity sequence, and let $(A_n) \in \mathcal{B}$ and $\varepsilon > 0$. Then the coset $(A_n)+\mathcal{K}$ is ε-invertible if and only if the coset $(A_{\eta(n)}) + \mathcal{K}_\eta$ is ε-invertible for a certain monotonically increasing sequence $\eta : \mathbb{N} \to \mathbb{N}$.

(b) For $(P_n A P_n) \in \mathcal{S}(\mathbb{N})$ and $\varepsilon > 0$,

$$\sigma^{(\varepsilon)}((P_n A P_n) + \mathcal{K}) = \sigma^{(\varepsilon)}(A + K(l^2(\mathbb{N}))).$$

PROOF. The first assertion can be checked as Theorem 6.7 and its Corollary 6.12 if one takes into account that

$$\|\mathbf{A} + \mathcal{K}\| = \|\pi^{\mathcal{K}}(R_\eta(\mathbf{A}) + \mathcal{K}_\eta)\| < \|R_\eta(\mathbf{A}) + \mathcal{K}_\eta\|$$

for each sequence $\mathbf{A} \in \mathcal{B}$. The second assertion follows immediately from Theorem 5.8. □

7.7. Pseudomodes

Let $(A_n) \in \mathcal{F}$. A point $\lambda \in \mathbb{C}$ is called an *asymptotically good pseudoeigenvalue* of (A_n) if

$$\|(A_n - \lambda I_n)^{-1}\| \to \infty \quad \text{as } n \to \infty$$

where we agree upon writing $\|(A_n - \lambda I_n)^{-1}\| = \infty$ if $A_n - \lambda I_n$ is not invertible. If λ is an asymptotically good pseudoeigenvalue of (A_n), then there is a sequence (x_n) of unit vectors $x_n \in \operatorname{im} P_n$ such that

$$\|(A_n - \lambda I_n)x_n\| \to 0 \quad \text{as } n \to \infty.$$

Each sequence (x_n) with this property is called an *asymptotically good pseudomode* of (A_n) at λ. This terminology has been introduced in [12] motivated by the papers [40, 56]; see also Chapter 12 in [13]. Notice that if λ is an asymptotically good pseudoeigenvalue of (A_n) and (x_n) an associated asymptotically good pseudomode, and if (G_n) is a sequence in $\mathcal{G}$, then λ is an asymptotically good pseudoeigenvalue of $(A_n + G_n)$ and (x_n) is an associated asymptotically good pseudomode of $(A_n + G_n)$. Thus, the asymptotically good pseudoeigenvalues and -modes of (A_n) depend only on the coset of (A_n) modulo $\mathcal{G}$. It is also clear that λ is an asymptotically good pseudoeigenvalue of (A_n) if and only if no subsequence of $(A_n - \lambda I_n)$ is stable. In particular, $\sigma(A)$ is contained in the set $\sigma_{agp}(\mathbf{A})$ of all asymptotically good pseudoeigenvalues of the finite sections sequence $\mathbf{A} = (P_n A P_n)$ (the existence of a stable subsequence of $(P_n A P_n)$ already implies the invertibility of A), and if $\mathbf{A} = (A_n) \in \mathcal{F}$ is a fractal sequence, then the set $\sigma_{agp}(\mathbf{A})$ coincides with the spectrum of the coset $\mathbf{A} + \mathcal{G}$ in $\mathcal{F}/\mathcal{G}$.

We are interested in the local behaviour of asymptotically good pseudomodes. Following Böttcher and Grudsky [**12**, **13**], we call a sequence (x_n) of unit vectors $x_n \in \operatorname{im} P_n$ *asymptotically strongly localized in the beginning part of the interval* $\{0, 1, \ldots, n-1\}$ if $\|P_{j_n} x_n\| \to 1$ as $n \to \infty$ for each sequence (j_n) of non-negative integers with $j_n \leq n-1$ and $j_n \to \infty$. The following two results are essentially Theorems 3.1 and 3.2 from [**12**].

PROPOSITION 7.18. *Let $B \in L(l^2(\mathbb{N}))$ be an operator for which the finite sections sequence $(P_n B P_n)$ is stable, and let m be a positive integer. Set $A := V_{-m} B$. Then a sequence (x_n) of unit vectors $x_n \in \operatorname{im} P_n$ is an asymptotically good pseudomode for $(P_n A P_n)$ at the asymptotically good pseudoeigenvalue 0 if and only if there are a bounded sequence (y_n) and a zero sequence (z_n) of vectors $y_n, z_n \in \operatorname{im} P_n$ such that $x_n = P_n B^{-1} P_m y_n + z_n$ for all $n \in \mathbb{N}$.*

PROOF. Notice first that the operator A is not invertible; hence, 0 is an asymptotically good pseudoeigenvalue of $(P_n A P_n)$. It is also clear that B is an invertible operator.

Let now $x_n = P_n B^{-1} P_m y_n + z_n$ be unit vectors with sequences (y_n) and (z_n) as described in the proposition. Then

$$\begin{aligned} \|P_n A P_n x_n\| &= \|P_n V_{-m} B P_n B^{-1} P_m y_n + P_n A P_n z_n\| \\ &\leq \|V_{-m} B P_n B^{-1} P_m\| \, \|y_n\| + \|A\| \, \|z_n\| \end{aligned}$$

whence $\|P_n A P_n x_n\| \to 0$ since the operators $V_{-m} B P_n B^{-1} P_m$ tend in the norm to $V_{-m} B B^{-1} P_m = 0$ as $n \to \infty$ since P_m is compact.

Conversely, let the sequence (x_n) of unit vectors $x_n \in \operatorname{im} P_n$ be an asymptotically good pseudomode for $(P_n A P_n)$. Write

$$P_n A P_n = P_n V_{-m} P_n B P_n + P_n V_{-m} Q_n B P_n =: A_n + B_n.$$

Since $\operatorname{im} A_n \subseteq \operatorname{im} P_{n-m}$ and $\operatorname{im} B_n \subseteq \operatorname{im}(P_n - P_{n-m})$, we have

$$\|P_n A P_n x_n\|^2 = \|A_n x_n + B_n x_n\|^2 = \|A_n x_n\|^2 + \|B_n x_n\|^2 \to 0$$

which implies that $\|A_n x_n\| \to 0$. Since $P_n V_m P_n V_{-m} P_n = P_n - P_m$ for $n > m$, we further conclude from $A_n x_n = P_n V_{-m} P_n B P_n x_n$ that

(7.18) $$P_n B P_n x_n = P_n V_m P_n A_n x_n + y_n$$

with certain vectors $y_n \in \operatorname{im} P_m$. It is also clear that the sequence (y_n) defined by (7.18) is bounded. Since $(P_n B P_n)$ is a stable sequence by assumption, we finally get from (7.18)

$$\begin{aligned} x_n &= (P_n B P_n)^{-1} y_n + (P_n B P_n)^{-1} P_n V_m P_n A_n x_n \\ &= P_n B^{-1} y_n + ((P_n B P_n)^{-1} - P_n B^{-1}) y_n + (P_n B P_n)^{-1} P_n V_m P_n A_n x_n \\ &=: P_n B^{-1} P_m y_n + z_n \end{aligned}$$

with a sequence (z_n) of vectors $z_n \in \operatorname{im} P_n$ tending to zero in the norm of $l^2(\mathbb{N})$. $\square$

THEOREM 7.19. *Let A, B and m be as in Proposition 7.18. Then each asymptotically good pseudomode for $(P_n A P_n)$ at the asymptotically good pseudoeigenvalue 0 is asymptotically strongly localized in the beginning part of the interval $\{0, 1, \ldots, n-1\}$.*

PROOF. Let (x_n) be an asymptotically good pseudomode for $(P_n A P_n)$ at the asymptotically good pseudoeigenvalue 0. By the preceding proposition, $x_n = P_n B^{-1} P_m y_n + z_n$ with a bounded sequence (y_n) and a zero sequence (z_n). Let (j_n) be a sequence of non-negative integers less than n which tends to infinity. Then

$$\begin{aligned} |1 - \|P_{j_n} x_n\|| &= |\|x_n\| - \|P_{j_n} x_n\|| \le \|(P_n - P_{j_n}) x_n\| \\ &\le \|(P_n - P_{j_n}) B^{-1} P_m y_n\| + \|(P_n - P_{j_n}) z_n\| \to 0 \end{aligned}$$

since $P_n - P_{j_n}$ tends strongly to zero and P_m is compact. $\square$

This theorem applies, e.g., to finite sections of Fredholm Toeplitz operators with continuous generating function which possess a positive index. Indeed, if $T(a)$ is an operator of this kind with index $m > 0$, then the Toeplitz operator $T(b)$ defined by $T(a) = V_{-m} T(b)$ has vanishing index. Thus, $T(b)$ is invertible due to Coburn's theorem, and the invertibility of $T(b)$ implies the stability of the finite sections method $(P_n T(b) P_n)$ (which is well known and follows easily from Theorem 2.12). Thus, by Theorem 7.19, each asymptotically good pseudomode for $(P_n T(a) P_n)$ at the asymptotically good pseudoeigenvalue 0 is asymptotically strongly localized in the beginning part of the interval $\{0, 1, \ldots, n-1\}$.

Notice that the continuity of a (hence, of b) is only needed to guarantee the implication

(7.19) $\qquad T(b)$ is invertible $\Rightarrow$ $(P_n T(b) P_n)$ is stable.

This implication holds for large classes of generating functions a such as piecewise continuous, piecewise quasicontinuous, and locally sectorial functions (but it does not hold for all functions $a \in L^\infty(\mathbb{T})$ as a famous example by Treil [58] shows). For an overview on related results including Treil's counterexample consult Chapter 7 of [14].

In order to get an analogous result for the finite sections of band-dominated operators with slowly oscillating functions, we need a further slight generalization of Theorem 7.19.

THEOREM 7.20. *Let $B \in L(l^2(\mathbb{N}))$ and $K \in K(l^2(\mathbb{N}))$ be such that the the finite sections sequence $(P_n(B+K)P_n)$ is stable. Let further $A := V_{-m} B$ with a positive integer m. Then each asymptotically good pseudomode for $(P_n A P_n)$ at the asymptotically good pseudoeigenvalue 0 is asymptotically strongly localized in the beginning part of the interval $\{0, 1, \ldots, n-1\}$.*

PROOF. Let (x_n) be an asymptotically good pseudomode for $(P_n A P_n)$ at the asymptotically good pseudoeigenvalue 0. As in the proof of Proposition 7.18 (identity (7.18)) we obtain
$$P_n B P_n x_n = P_m B P_n x_n + z_n$$
with a sequence (z_n) tending to zero. Thus,
$$P_n (B+K) P_n x_n = P_m B P_n x_n + P_n K P_n x_n + z_n.$$
Since $(B_n) := (P_n(B+K)P_n)$ is a stable sequence, we further get for all sufficiently large n
$$x_n = B_n^{-1} P_m B P_n x_n + B_n^{-1} P_n K P_n x_n + B_n^{-1} z_n.$$
Let now (j_n) be a sequence of non-negative integers $j_n \leq n$ which tends to infinity. Then
$$\begin{aligned} |1 - \|P_{j_n} x_n\|| &= |\|x_n\| - \|P_{j_n} x_n\|| \leq \|(P_n - P_{j_n}) x_n\| \\ &\leq \|(P_n - P_{j_n}) B_n^{-1} P_m B P_n x_n\| \\ &\quad + \|(P_n - P_{j_n}) B_n^{-1} P_n K P_n x_n\| + \|(P_n - P_{j_n}) z_n\| \\ &\leq \|(P_n - P_{j_n}) B_n^{-1} P_m\| \|B\| + \|(P_n - P_{j_n}) B_n^{-1} P_n K\| + \|z_n\|, \end{aligned}$$
and the right-hand side of this estimate tends to zero since $P_n - P_{j_n}$ tends strongly to zero and since P_m and K are compact. □

COROLLARY 7.21. *Let $A \in \mathcal{A}_{SO}(\mathbb{N})$ be a Fredholm operator with positive Fredholm index. Then every asymptotically good pseudomode for $(P_n A P_n)$ at the asymptotically good pseudoeigenvalue 0 is asymptotically strongly localized in the beginning part of $\{0, 1, \ldots, n-1\}$.*

PROOF. Write $A = V_{-m} B$ with an Fredholm operator $B \in \mathcal{A}_{SO}(\mathbb{N})$ with Fredholm index zero, and choose a compact operator K such that $B + K$ is invertible. Since the band-dominated operator $B + K$ has slowly oscillating coefficients, Theorem 2.10 ensures the stability of the sequence $(P_n(B+K)P_n)$. □

We conclude this section by a short discussion of pseudomodes of general stably regularizable sequences.

PROPOSITION 7.22. *Let $(A_n) \in \mathcal{F}$ be a stably regularizable sequence for which the point 0 is an asymptotically good pseudoeigenvalue, and let $(\Pi_n) + \mathcal{G}$ be the Moore-Penrose projection of the coset $(A_n) + \mathcal{G}$. Then a sequence (x_n) of unit vectors $x_n \in \operatorname{im} P_n$ is an asymptotically good pseudomode for (A_n) at 0 if and only if*

(7.20) $$\|x_n - \Pi_n x_n\| \to 0 \quad as\ n \to \infty.$$

PROOF. Recall that $\|A_n \Pi_n\| \to 0$ and that the sequence $(A_n^* A_n + \Pi_n)$ is stable, and set $C := \sup \|(A_n^* A_n + \Pi_n)^{-1}\|$. If (x_n) is a sequence of unit vectors subject to condition (7.20), then
$$\begin{aligned} \|A_n x_n\| &\leq \|A_n(x_n - \Pi_n x_n)\| + \|A_n \Pi_n x_n\| \\ &\leq \|A_n\| \|x_n - \Pi_n x_n\| + \|A_n \Pi_n\| \to 0. \end{aligned}$$

Hence, (x_n) is an asymptotically good pseudomode for (A_n) at 0. Conversely, for each asymptotically good pseudomode (x_n) for (A_n) at 0,

$$\begin{aligned} \|x_n - \Pi_n x_n\| &= \|(A_n^* A_n + \Pi_n)^{-1}(A_n^* A_n + \Pi_n)(x_n - \Pi_n x_n)\| \\ &\leq C\|(A_n^* A_n + \Pi_n)(x_n - \Pi_n x_n)\| \\ &\leq C\|A_n\|\|A_n x_n\| + C\|A_n\|\|A_n \Pi_n\| + \|\Pi_n(I - \Pi_n)\| \to 0. \end{aligned}$$

Thus, (x_n) satisfies (7.20). □

If, in addition to the hypotheses of the preceding proposition, (Π_n) is of the form $(P_n \Pi P_n + G_n)$ with a compact projection Π and a sequence (G_n) in $\mathcal{G}$, then one has $\|(P_n - P_{j_n})\Pi P_n\| \to 0$ for every sequence (j_n) of non-negative integers $j_n \leq n$ which tends to infinity. Under these assumptions, every asymptotically good pseudomode (x_n) for $(P_n A P_n)$ at 0 is asymptotically strongly localized in the beginning part of $\{0, 1, \ldots, n-1\}$. Indeed,

$$\begin{aligned} |1 - \|P_{j_n} x_n\|| &\leq \|(P_n - P_{j_n}) x_n\| \\ &\leq \|(P_n - P_{j_n})\Pi_n x_n\| + \|(P_n - P_{j_n})(I - \Pi_n) x_n\| \\ &\leq \|(P_n - P_{j_n})\Pi_n\| + \|(I - \Pi_n) x_n\| \to 0 \end{aligned}$$

by Proposition 7.22. It would be certainly of interest to study stably regularizable sequences the Moore-Penrose projection of which is of the form $(P_n \Pi P_n + G_n)$ with a compact projection Π and a sequence (G_n) in $\mathcal{G}$.

7.8. Determinants

Let A be a band-dominated operator on $l^2(\mathbb{N})$ for which the finite sections method $(P_n A P_n)$ is stable. Then the matrices $P_n A P_n$ are invertible for n large enough, and it makes sense to consider the sequence

$$(7.21) \qquad n \mapsto \frac{\det(P_{n-1} A P_{n-1})}{\det(P_n A P_n)}.$$

In case $A = T(a)$ is an invertible Toeplitz operator, the sequence (7.21) converges, and its limit can be expressed in terms of the generating function a. This is the classical Szegö limit theorem; a nice introduction to this topic is [**15**]. For a general band-dominated operator A, one cannot expect convergence of (7.21) as the band operator

$$A := \operatorname{diag}\left(\begin{pmatrix} 2 & 1 \\ 1 & 2 \end{pmatrix}, \begin{pmatrix} 2 & 1 \\ 1 & 2 \end{pmatrix}, \begin{pmatrix} 2 & 1 \\ 1 & 2 \end{pmatrix}, \ldots \right)$$

shows. In this case we denote by $\omega(A)$ the set of all partial limits of the sequence (7.21). It turns out that this set can be described via limit operators.

THEOREM 7.23. *Let $A \in L(l^2(\mathbb{N}))$ be a band-dominated operator for which the finite sections sequence $(P_n A P_n)$ is stable. Then*

$$(7.22) \qquad \omega(A) = \{P_1 (J Q A_h Q J|_{l^2(\mathbb{N})})^{-1} P_1 : A_h \in \sigma_+(A)\}.$$

For the notations, see Section 2.4. The identity (7.22) has to be read as follows: $P_1 (J Q A_h Q J|_{l^2(\mathbb{N})})^{-1} P_1$ can be understood as an 1×1-matrix, and we identify this matrix with its only entry, which is a complex number.

PROOF. For n a positive integer, let
$$W_n : l^2(\mathbb{N}) \to l^2(\mathbb{N}), \quad (x_0, x_1, \ldots) \mapsto (x_{n-1}, x_{n-2}, \ldots, x_0, 0, 0, \ldots).$$
If the finite sections method $(P_n A P_n)$ is stable, then the operators $W_n A W_n$, considered as acting on $\operatorname{im} W_n = \operatorname{im} P_n$, are invertible for large n, and
$$\frac{\det(P_{n-1} A P_{n-1})}{\det(P_n A P_n)} = \frac{\det(W_{n-1} A W_{n-1})}{\det(W_n A W_n)} =: \beta_n.$$
By Cramer's rule, β_n equals the first component of the solution $x^{(n)}$ to the equation
$$W_n A W_n x^{(n)} = (1, 0, 0, \ldots, 0)^T.$$
Let now $\alpha \in \omega(A)$, and let $h : \mathbb{N} \to \mathbb{N}$ be a sequence tending to infinity such that $\alpha = \lim \beta_{h(n)}$. Since A is band-dominated, there is a subsequence g of h such that the limit operator
$$A_g = \text{s-lim}\, U_{-g(n)} A U_{g(n)} \in L(l^2(\mathbb{Z}))$$
exists. Then also the strong limit
$$\text{s-lim}\, J U_{-g(n)} P_{g(n)} A P_{g(n)} U_{g(n)} J \in L(l^2(\mathbb{N}))$$
exists and is equal to $JQ A_g QJ$. Since $JU_{-n} P_n = W_n$ and $P_n U_n J = W_n$, this shows that the strong limit
$$\text{s-lim}\, W_{g(n)} A W_{g(n)} \quad \text{exists and is equal to} \quad JQ A_g QJ \in L(l^2(\mathbb{N})).$$
So we can consider the sequence $(W_{g(n)} A W_{g(n)})_{n \in \mathbb{N}}$ as a stable and convergent approximation method for the operator $JQ A_g QJ$. In particular, the solutions $x^{(n)}$ to the equation

(7.23) $$W_{g(n)} A W_{g(n)} x^{(n)} = (1, 0, 0, \ldots, 0)^T$$

converge in the norm of $l^2(\mathbb{N})$ to the solution x to the equation

(7.24) $$JQ A_g J x = (1, 0, 0, \ldots)^T.$$

Thus, the first component $\beta_{g(n)}$ of the solution $x^{(n)}$ to equation (7.23) converges to the first component of the solution x to equation (7.24). Since the latter one is equal to
$$P_1 x = P_1 (JQ A_g QJ)^{-1} P_1,$$
we arrive at $\alpha = P_1 (JQ A_g QJ)^{-1} P_1$. This is the inclusion $\subseteq$ in (7.22). The reverse inclusion can be proved by similar arguments. □

In case $A = T(a)$ is a Toeplitz operator, the sequence $(W_n T(a) W_n)$ converges strongly, and its strong limit is the Toeplitz operator $T(\tilde{a})$ with $\tilde{a}(t) := a(1/t)$. Hence, in this case, $\omega(T(a))$ is the singleton $\{P_1 T(\tilde{a})^{-1} P_1\}$. Consequently, the sequence (β_n) converges, and its limit is $P_1 T(\tilde{a})^{-1} P_1$, considered as a complex number. Under suitable assumptions for a (e.g., belonging to the Wiener algebra or being locally sectorial) one can identify the number $P_1 T(\tilde{a})^{-1} P_1$ with $1/\exp(\log a)_0$ with b_0 referring to the 0th Fourier coefficient of the function b (details can be found in Section 5.4 of [**15**], for example).

In case $A \in L(l^2(\mathbb{N}))$ is a band-dominated operator with slowly oscillating coefficients, all limit operators of A are shift invariant; hence, all partial limits in $\omega(A)$ are of the form $P_1 T(\tilde{a}_h)^{-1} P_1$ with certain continuous functions a_h. If, moreover, $A = \sum a_k V_k$ satisfies the Wiener condition $\sum \|a_k\|_\infty < \infty$, then all functions a_h belong to the Wiener algebra, and one has again $P_1 T(\tilde{a}_h)^{-1} P_1 = 1/\exp(\log a_h)_0$.

Bibliography

[1] G. R. Allan. Ideals of vector valued functions. *Proc. Lond. Math. Soc.*, 18(3):193 – 216, 1968.

[2] W. Arveson. Improper filtrations for C^*-algebras: Spectra of unilateral tridiagonal operators. *Acta Sci. Math. (Szeged)*, 57:11 – 24, 1993.

[3] W. Arveson. C^*-algebras and numerical linear algebra. *J. Funct. Anal*, 122:333 – 360, 1994.

[4] W. Arveson. The role of C^*-algebras in infinite dimensional numerical linear algebra. *Contemp. Math.*, 167:115 – 129, 1994.

[5] A. Avila and S. Jitomirskaya. The ten martini problem. *arXiv:math*, DS/0503363, March 2005.

[6] A. Avila and S. Jitomirskaya. Solving the ten Martini problem. In *Mathematical Physics of Quantum Mechanics*, volume 690 of *Lecture Notes Phys*, pages 5 – 16. Springer, Berlin, 2006.

[7] M. Bastos, A. Bravo, and I. Karlovich. Convolution type operators with symbols generated by slowly oscillating and piecewise continuous matrix functions. In I. Gohberg, A.F. dos Santos, F.-O. Speck, S. Teixera, and W.L. Wendland, editors, *Operator Theoretical Methods and Applications to Mathematical Physics*, volume 147 of *Operator Theory: Adv. and Appl.*, pages 151 – 174. Birkhäuser Verlag, Basel, Boston, Berlin, 2004.

[8] F. P. Boca. *Rotation C^*-Algebras and Almost Mathieu Operators*, volume 1 of *Theta Series in Advanced Mathematics*. The Theta Foundation, Bucharest, 2001.

[9] A. Böttcher. Pseudospectra and singular values of large convolution operators. *J. Integral Equations Appl.*, 6(3):267 – 301, 1994.

[10] A. Böttcher, A. V. Chithra, and M. N. N. Namboodiri. Approximation of approximation numbers by truncation. *Integral Equations Oper. Theory*, 39(4):387 – 395, 2001.

[11] A. Böttcher and S. M. Grudsky. *Toeplitz matrices, Asymptotic Linear Algebra and Functional Analysis*. Hindustan Book Agency, New Delhi, 2000.

[12] A. Böttcher and S. M. Grudsky. Asymptotically good pseudomodes for Toeplitz matrices and Wiener-Hopf operators. In I. Gohberg, A.F. dos Santos, F.-O. Speck, S. Teixera, and W.L. Wendland, editors, *Operator Theoretical Methods and Applications to Mathematical Physics*, volume 147 of *Operator Theory: Adv. and Appl.*, pages 175 – 188. Birkhäuser Verlag, Basel, 2004.

[13] A. Böttcher and S. M. Grudsky. *Spectral Properties of Banded Toeplitz Matrices*. siam, Philadelphia, 2005.

[14] A. Böttcher and B. Silbermann. *Analysis of Toeplitz Operators*. Akademie-Verlag, Berlin and Springer-Verlag, Berlin, Heidelberg, New York, 1989/1990.

[15] A. Böttcher and B. Silbermann. *Introduction to Large Truncated Toeplitz Matrices*. Springer-Verlag, Berlin, Heidelberg, 1999.

[16] R. G. Douglas. *Banach Algebra Techniques in Operator Theory*. Academic Press, New York, 1972.

[17] T. Ehrhardt, S. Roch, and B. Silbermann. Szegö limit theorems for operators with almost periodic coefficients II. In preparation.

[18] I. Gohberg and I. Feldman. *Convolution Equations and Projection Methods for Their Solution*. Nauka, Moskva, 1971. Russian, Engl. transl.: Amer. Math. Soc. Transl. of Math. Monographs, Vol. 41, Providence, Rhode Island, 1974.

[19] I. Gohberg, S. Goldberg, and M. A. Kaashoek. *Classes of Linear Operators, Volume I*. Birkhäuser Verlag, Basel, Boston, Berlin, 1990.

[20] I. Gohberg and M. A. Kaashoek. Projection method for Block Toeplitz operators with operator-valued symbols. In *Toeplitz Operators and Related Topics*, volume 71 of *Operator Theory: Adv. and Appl.*, pages 79 – 104. Birkhäuser Verlag, Basel, Boston, Berlin, 1994.

[21] I. Gohberg and N. Krupnik. *One-dimensional Linear Singular Integral Operators, Volume I.* Birkhäuser Verlag, Basel, Boston, Stuttgart, 1992.

[22] R. Hagen, S. Roch, and B. Silbermann. *Spectral Theory of Approximation Methods for Convolution Equations.* Birkhäuser Verlag, Basel, Boston, Berlin, 1995.

[23] R. Hagen, S. Roch, and B. Silbermann. C^*-*Algebras and Numerical Analysis.* Marcel Dekker, Inc., New York, Basel, 2001.

[24] N. Higson and J. Roe. *Analytic K-Homology.* Oxford University Press, Oxford, 2000.

[25] M. Lindner, V. S. Rabinovich, and S. Roch. Finite sections of band operators with slowly oscillating coefficients. *Linear Alg. Appl.*, 390:19 – 26, 2004.

[26] R. H. Moore and M. Z. Nashed. Approximation of generalized inverses of linear operators. *SIAM J. Appl. Math.*, 27(1):1 – 16, 1974.

[27] G. K. Pedersen. C^*-*Algebras and Their Automorphism Groups.* Academic Press, London, New York, San Francisco, 1979.

[28] A. Pietsch. *Theorie der Operatorenideale.* Wissenschaftliche Beiträge Friedrich-Schiller-Univ., Jena, 1972.

[29] A. Pietsch. *Operator Ideals.* Deutscher Verlag der Wissenschaften, Berlin, 1978.

[30] S. Prössdorf and B. Silbermann. *Projektionsverfahren und die näherungsweise Lösung singulärer Gleichungen.* B. G. Teubner Verlagsgesellschaft, Leipzig, 1977.

[31] S. Prössdorf and B. Silbermann. *Numerical Analysis for Integral and Related Operator Equations.* Akademie-Verlag, Berlin and Birkhäuser Verlag, Basel, Boston, Stuttgart, 1991.

[32] V. S. Rabinovich and S. Roch. Local theory of the Fredholmness of band-dominated operators with slowly oscillating coefficients. In A. Böttcher, I. Gohberg, and P. Junghanns, editors, *Toeplitz Matrices, Convolution Operators, and Integral Equations*, volume 135 of *Operator Theory: Adv. and Appl.*, pages 267 – 291. Birkhäuser Verlag, Basel, Boston, Berlin, 2002.

[33] V. S. Rabinovich and S. Roch. Algebras of approximation sequences: Spectral and pseudospectral approximation of band-dominated operators. In S. H. Kulkarni and M. N. N. Namboodiri, editors, *Proc. Int. Workshop Linear Algebra, Numerical Functional Analysis and Wavelet Analysis, Cochin (India) 2001*, pages 167 – 188. Allied Publ. Private Limited, New Delhi, 2003.

[34] V. S. Rabinovich, S. Roch, and J. Roe. Fredholm indices of band-dominated operators. *Integral Equations Oper. Theory*, 49(2):221 – 238, 2004.

[35] V. S. Rabinovich, S. Roch, and B. Silbermann. Fredholm theory and finite section method for band-dominated operators. *Integral Equations Oper. Theory*, 30(4):452 – 495, 1998.

[36] V. S. Rabinovich, S. Roch, and B. Silbermann. Algebras of approximation sequences: Finite sections of band-dominated operators. *Acta Appl. Math.*, 65:315 – 332, 2001.

[37] V. S. Rabinovich, S. Roch, and B. Silbermann. *Limit Operators and Their Applications in Operator Theory*, volume 150 of *Operator Theory: Adv. and Appl.* Birkhäuser Verlag, Basel, Boston, Berlin, 2004.

[38] V. S. Rabinovich, S. Roch, and B. Silbermann. On finite sections of band-dominated operators. Submitted to Proc. WOAT Lisbon, 2006.

[39] V. S. Rabinovich, S. Roch, and B. Silbermann. Finite sections of band-dominated operators with almost periodic coefficients. Preprint 2398 TU Darmstadt, 26 pages, Mai 2005. To appear in Operator Theory: Adv. and Appl.

[40] L. Reichel and L. N. Trefethen. Eigenvalues and pseudo-eigenvalues of Toeplitz matrices. *Linear Algebra Appl.*, 162:153 – 185, 1992.

[41] S. Roch. Algebras of approximation sequences: Fractality. In J. Elschner, I. Gohberg, and B. Silbermann, editors, *Problems and Methods in Mathematical Physics*, volume 121 of *Operator Theory: Adv. and Appl.*, pages 471 – 497. Birkhäuser Verlag, Basel, 2001.

[42] S. Roch. Algebras of approximation sequences: Fredholm theory in fractal algebras. *Studia Math.*, 150(1):53 – 77, 2002.

[43] S. Roch. Algebras of approximation sequences: Fredholmness. *J. Oper. Theory*, 48:121 – 149, 2002.

[44] S. Roch. Band-dominated operators on l^p-spaces: Fredholm indices and finite sections. *Acta Sci. Math. (Szeged)*, 70(3/4):783 – 797, 2004.

[45] S. Roch and B. Silbermann. Szegö limit theorems for operators with almost periodic coefficients. To appear in Operators and Matrices.

[46] S. Roch and B. Silbermann. C^*-algebra techniques in numerical analysis. *J. Oper. Theory*, 35(2):241 – 280, 1996.

[47] S. Roch and B. Silbermann. Index calculus for approximation methods and singular value decomposition. *J. Math. Anal. Appl.*, 225:401 – 426, 1998.

[48] S. Roch and B. Silbermann. Continuity of generalized inverses in Banach algebras. *Studia Math.*, 136(3):197 – 227, 1999.

[49] J. Roe. *Index Theory, Coarse Geometry, and Topology of Manifolds*, volume 90 of *CBMS*. Amer. Math. Soc., Providence, R. I., 1996.

[50] J. Roe. *Lectures on Coarse geometry*, volume 31 of *Univ. Lecture Series*. Amer. Math. Soc., Providence, R. I., 2003.

[51] J. Roe. Band-dominated Fredholm operators on discrete groups. *Integral Equations Oper. Theory*, 51(3):411 – 416, 2005.

[52] M. Rørdam, F. Larsen, and N. J. Laustsen. *An Introduction to K-Theory for C^*-Algebras*, volume 49 of *London Math Soc. Student Texts*. Cambridge University Press, 2000.

[53] B. Silbermann. Asymptotic Moore – Penrose inversion of Toeplitz operators. *Linear Algebra Appl.*, 256:219 – 234, 1997.

[54] B. Silbermann. How to compute the partial indices of a regular and smooth matrix-valued function. In *Factorization, singular operators and related problems*, pages 291 – 300. Kluwer Akad. Publ., Dordrecht, 2003.

[55] B. Silbermann. Modified finite sections for Toeplitz operators and their singular values. *SIAM J. Matrix Anal. Appl.*, 24(3):678 – 692, 2003.

[56] L. N. Trefethen and S. J. Chapman. Wave packet pseudomodes of twisted Toeplitz matrices. *Comm. Pure Appl. Math.*, 57(9):1233 – 1264, 2004.

[57] L. N. Trefethen and M. Embree. *Spectra and Pseudospectra. The Behavior of Nonnormal Matrices and Operators*. Princeton Univ. Press, Princeton, Oxford, 2005.

[58] S. R. Treil. Invertibility of Toeplitz operators does not imply applicability of the finite section method . *Dokl. Akad. Nauk SSSR*, 292(3):563 – 567, 1987. (Russian).

Index

$N(\lambda)$, 71
U_k, 2
V_k, 3
$\Sigma_k(A)$, 32
α-number, 48
$\mathcal{A}(\mathbb{Z})$, 1
$\mathcal{A}(\mathbb{Z}_\pm)$, 2
$\mathcal{F}(\mathbb{N})$, 8
$\mathcal{F}(\mathbb{Z})$, 8
$\mathcal{G}(\mathbb{N})$, 8
$\mathcal{G}(\mathbb{Z})$, 8
$\mathcal{K}$, 30
$\mathcal{K}(\mathbb{Z})$, 45
$\mathcal{K}_0$, 30
$\mathcal{K}_0(\mathbb{Z})$, 18
$\mathcal{S}(\mathbb{N})$, 14
$\mathcal{S}(\mathbb{Z})$, 18
$\mathcal{S}_{P\mathbb{C}}(\mathbb{Z})$, 62
δ, 8
$\omega(A)$, 80
$\sigma(A)$, 24
$\sigma_\pm(A)$, 7
$\sigma_k(A)$, 32
$\sigma_{agp}(\mathbf{A})$, 77
$\sigma_{op,1}(\mathbf{A})$, 11
$\sigma_{op}(A)$, 6
$\sigma_{sing}(A)$, 25
ε-invertibility, 73
   essential, 73
ε-pseudospectrum, 73
   essential, 73
$a^\dagger$, 23

algebra
   $\mathcal{J}$-fractal, 59
   essentially fractal, 61
   fractal, 61
   of matrix sequences, 8
   of the finite sections method, 14
approximation number, 32, 64
Arveson dichotomy, 67

center, 35

dimension function, 8
discretization, 65
   rank-preserving, 65

element
   centrally compact, 35
   of central rank one, 35
   of finite central rank, 35

finite sections method, 5

homomorphism
   fractal, 57

ideal
   quasicommutator, 15
inverse
   Moore-Penrose, 23

Moore-Penrose invertibility, 23
Moore-Penrose projection, 25

norm
   esential, 32

operator
   Almost Mathieu, 1
   discrete Schrödinger, 1
   Fredholm, 5
operator spectrum, 6
   of a sequence, 11

point
   essential, 67
   transient, 67
pseudoeigenvalue
   asymptotically good, 76
pseudomode
   asymptotically good, 76

rank
   central, 35
   essential, 30

sequence
   compact, 30
   compact in $\mathcal{F}(\mathbb{Z})$, 45

not normally solvable, 62
  of s-numbers, 64
  of finite essential rank, 30
  of rank one matrices, 30
  stable, 5
  stably regularizable, 23
singular value decomposition, 32
spectrum
  essential, 50
splitting property, 25
symbol, 7
Szegö limit theorem, 80

theorem
  by Kozak, 8

Editorial Information

To be published in the *Memoirs*, a paper must be correct, new, nontrivial, and significant. Further, it must be well written and of interest to a substantial number of mathematicians. Piecemeal results, such as an inconclusive step toward an unproved major theorem or a minor variation on a known result, are in general not acceptable for publication.

Papers appearing in *Memoirs* are generally at least 80 and not more than 200 published pages in length. Papers less than 80 or more than 200 published pages require the approval of the Managing Editor of the Transactions/Memoirs Editorial Board.

As of September 30, 2007, the backlog for this journal was approximately 14 volumes. This estimate is the result of dividing the number of manuscripts for this journal in the Providence office that have not yet gone to the printer on the above date by the average number of monographs per volume over the previous twelve months, reduced by the number of volumes published in four months (the time necessary for preparing a volume for the printer). (There are 6 volumes per year, each usually containing at least 4 numbers.)

A Consent to Publish and Copyright Agreement is required before a paper will be published in the *Memoirs*. After a paper is accepted for publication, the Providence office will send a Consent to Publish and Copyright Agreement to all authors of the paper. By submitting a paper to the *Memoirs*, authors certify that the results have not been submitted to nor are they under consideration for publication by another journal, conference proceedings, or similar publication.

Information for Authors

Memoirs are printed from camera copy fully prepared by the author. This means that the finished book will look exactly like the copy submitted.

Initial submission. The AMS uses Centralized Manuscript Processing for initial submissions. Authors should submit a PDF file using the Initial Manuscript Submission form found at www.ams.org/cgi-bin/peertrack/submission.pl, or send one copy of the manuscript to the following address: Centralized Manuscript Processing, MEMOIRS OF THE AMS, 201 Charles Street, Providence, RI 02904-2294 USA. If a paper copy is being forwarded to the AMS, indicate that it is for it Memoirs and include the name of the corresponding author, contact information such as email address or mailing address, and the name of an appropriate Editor to review the paper (see the list of Editors below).

The paper must contain a *descriptive title* and an *abstract* that summarizes the article in language suitable for workers in the general field (algebra, analysis, etc.). The *descriptive title* should be short, but informative; useless or vague phrases such as "some remarks about" or "concerning" should be avoided. The *abstract* should be at least one complete sentence, and at most 300 words. Included with the footnotes to the paper should be the 2000 *Mathematics Subject Classification* representing the primary and secondary subjects of the article. The classifications are accessible from www.ams.org/msc/. The list of classifications is also available in print starting with the 1999 annual index of *Mathematical Reviews*. The Mathematics Subject Classification footnote may be followed by a list of *key words and phrases* describing the subject matter of the article and taken from it. Journal abbreviations used in bibliographies are listed in the latest *Mathematical Reviews* annual index. The series abbreviations are also accessible from www.ams.org/publications/. To help in preparing and verifying references, the AMS offers MR Lookup, a Reference Tool for Linking, at www.ams.org/mrlookup/.

Electronically prepared manuscripts. The AMS encourages electronically prepared manuscripts, with a strong preference for $\mathcal{A}_{\mathcal{M}}\mathcal{S}$-LaTeX. To this end, the Society has prepared $\mathcal{A}_{\mathcal{M}}\mathcal{S}$-LaTeX author packages for each AMS publication. Author packages include instructions for preparing electronic manuscripts, samples, and a style file that generates the particular design specifications of that publication series. Though $\mathcal{A}_{\mathcal{M}}\mathcal{S}$-LaTeX is the highly preferred format of TeX, author packages are also available in $\mathcal{A}_{\mathcal{M}}\mathcal{S}$-TeX.

Authors may retrieve an author package from the AMS website starting from www.ams.org/tex/ or via FTP to ftp.ams.org (login as anonymous, enter username as password, and type cd pub/author-info). The *AMS Author Handbook* and the *Instruction Manual* are available in PDF format following the author packages link from www.ams.org/tex/. The author package can also be obtained free of charge by sending email to tech-support@ams.org (Internet) or from the Publication Division, American Mathematical Society, 201 Charles St., Providence, RI 02904-2294, USA. When requesting an author package, please specify $\mathcal{A}_{\mathcal{M}}\mathcal{S}$-LaTeX or $\mathcal{A}_{\mathcal{M}}\mathcal{S}$-TeX and the publication in which your paper will appear. Please be sure to include your complete mailing address.

After acceptance. The final version of the electronic file should be sent to the Providence office (this includes any TeX source file, any graphics files, and the DVI or PostScript file) immediately after the paper has been accepted for publication.

Before sending the source file, be sure you have proofread your paper carefully. The files you send must be the EXACT files used to generate the proof copy that was accepted for publication. For all publications, authors are required to send a printed copy of their paper, which exactly matches the copy approved for publication, along with any graphics that will appear in the paper.

Accepted electronically prepared files can be submitted via the web at www.ams.org/submit-book-journal/, sent via FTP, or sent on CD-Rom or diskette to the Electronic Prepress Department, American Mathematical Society, 201 Charles Street, Providence, RI 02904-2294 USA. TeX source files, DVI files, and PostScript files can be transferred over the Internet by FTP to the Internet node ftp.ams.org (130.44.1.100). When sending a manuscript electronically via CD-Rom or diskette, please be sure to include a message identifying the paper as a Memoir.

Electronically prepared manuscripts can also be sent via email to pub-submit@ams.org (Internet). In order to send files via email, they must be encoded properly. (DVI files are binary and PostScript files tend to be very large.)

Electronic graphics. Comprehensive instructions on preparing graphics are available at www.ams.org/jourhtml/. A few of the major requirements are given here.

Submit files for graphics as EPS (Encapsulated PostScript) files. This includes graphics originated via a graphics application as well as scanned photographs or other computer-generated images. If this is not possible, TIFF files are acceptable as long as they can be opened in Adobe Photoshop or Illustrator. No matter what method was used to produce the graphic, it is necessary to provide a paper copy to the AMS.

Authors using graphics packages for the creation of electronic art should also avoid the use of any lines thinner than 0.5 points in width. Many graphics packages allow the user to specify a "hairline" for a very thin line. Hairlines often look acceptable when proofed on a typical laser printer. However, when produced on a high-resolution laser imagesetter, hairlines become nearly invisible and will be lost entirely in the final printing process.

Screens should be set to values between 15% and 85%. Screens which fall outside of this range are too light or too dark to print correctly. Variations of screens within a graphic should be no less than 10%.

Inquiries. Any inquiries concerning a paper that has been accepted for publication should be sent to memo-query@ams.org or directly to the Electronic Prepress Department, American Mathematical Society, 201 Charles St., Providence, RI 02904-2294 USA.

Editors

This journal is designed particularly for long research papers, normally at least 80 pages in length, and groups of cognate papers in pure and applied mathematics. Papers intended for publication in the *Memoirs* should be addressed to one of the following editors. The AMS uses Centralized Manuscript Processing for initial submissions to AMS journals. Authors should follow instructions listed on the Initial Submission page found at www.ams.org/memo/memosubmit.html.

Algebra to ALEXANDER KLESHCHEV, Department of Mathematics, University of Oregon, Eugene, OR 97403-1222; email: ams@noether.uoregon.edu

Algebraic geometry and its application to MINA TEICHER, Emmy Noether Research Institute for Mathematics, Bar-Ilan University, Ramat-Gan 52900, Israel; email: teicher@macs.biu.ac.il

Algebraic geometry to DAN ABRAMOVICH, Department of Mathematics, Brown University, Box 1917, Providence, RI 02912; email: amsedit@math.brown.edu

Algebraic number theory to V. KUMAR MURTY, Department of Mathematics, University of Toronto, 100 St. George Street, Toronto, ON M5S 1A1, Canada; email: murty@math.toronto.edu

Algebraic topology to ALEJANDRO ADEM, Department of Mathematics, University of British Columbia, Room 121, 1984 Mathematics Road, Vancouver, British Columbia, Canada V6T 1Z2; email: adem@math.ubc.ca

Combinatorics to JOHN R. STEMBRIDGE, Department of Mathematics, University of Michigan, Ann Arbor, Michigan 48109-1109; email: FRS@umich.edu

Complex analysis and harmonic analysis to ALEXANDER NAGEL, Department of Mathematics, University of Wisconsin, 480 Lincoln Drive, Madison, WI 53706-1313; email: nagel@math.wisc.edu

Differential geometry and global analysis to LISA C. JEFFREY, Department of Mathematics, University of Toronto, 100 St. George St., Toronto, ON Canada M5S 3G3; email: jeffrey@math.toronto.edu

Dynamical systems and ergodic theory to AMIE WILKINSON, Department of Mathematics, Northwestern University, 2033 Sheridan Road, Evanston, IL 60208-2730; email: transactions@math.northwestern.edu

Functional analysis and operator algebras to DIMITRI SHLYAKHTENKO, Department of Mathematics, University of California, Los Angeles, CA 90095; email: shlyakht@math.ucla.edu

Geometric analysis to WILLIAM P. MINICOZZI II, Department of Mathematics, Johns Hopkins University, 3400 N. Charles St., Baltimore, MD 21218; email: trans@math.jhu.edu

Geometric analysis to MLADEN BESTVINA, Department of Mathematics, University of Utah, 155 South 1400 East, JWB 233, Salt Lake City, Utah 84112-0090; email: bestvina@math.utah.edu

Harmonic analysis, representation theory, and Lie theory to ROBERT J. STANTON, Department of Mathematics, The Ohio State University, 231 West 18th Avenue, Columbus, OH 43210-1174; email: stanton@math.ohio-state.edu

Logic to STEFFEN LEMPP, Department of Mathematics, University of Wisconsin, 480 Lincoln Drive, Madison, Wisconsin 53706-1388; email: lempp@math.wisc.edu

Partial differential equations to GUSTAVO PONCE, Department of Mathematics, South Hall, Room 6607, University of California, Santa Barbara, CA 93106; email: ponce@math.ucsb.edu

Partial differential equations and dynamical systems to PETER POLACIK, School of Mathematics, University of Minnesota, Minneapolis, MN 55455; email: polacik@math.umn.edu

Probability and statistics to KRZYSZTOF BURDZY, Department of Mathematics, University of Washington, Box 354350, Seattle, Washington 98195-4350; email: burdzy@math.washington.edu

Real analysis and partial differential equations to DANIEL TATARU, Department of Mathematics, University of California, Berkeley, Berkeley, CA 94720; email: tataru@math.berkeley.edu

All other communications to the editors should be addressed to the Managing Editor, ROBERT GURALNICK, Department of Mathematics, University of Southern California, Los Angeles, CA 90089-1113; email: guralnic@math.usc.edu.

Titles in This Series

895 **Steffen Roch,** Finite sections of band-dominated operators, 2008

894 **Martin Dindoš,** Hardy spaces and potential theory on C^1 domains in Riemannian manifolds, 2008

893 **Tadeusz Iwaniec and Gaven Martin,** The Beltrami Equation, 2008

892 **Jim Agler, John Harland, and Benjamin J. Raphael,** Classical function theory, operator dilation theory, and machine computation on multiply-connected domains, 2008

891 **John H. Hubbard and Peter Papadopol,** Newton's method applied to two quadratic equations in $\mathbb{C}^2$ viewed as a global dynamical system, 2008

890 **Steven Dale Cutkosky,** Toroidalization of dominant morphisms of 3-folds, 2007

889 **Michael Sever,** Distribution solutions of nonlinear systems of conservation laws, 2007

888 **Roger Chalkley,** Basic global relative invariants for nonlinear differential equations, 2007

887 **Charlotte Wahl,** Noncommutative Maslov index and eta-forms, 2007

886 **Robert M. Guralnick and John Shareshian,** Symmetric and alternating groups as monodromy groups of Riemann surfaces I: Generic covers and covers with many branch points, 2007

885 **Jae Choon Cha,** The structure of the rational concordance group of knots, 2007

884 **Dan Haran, Moshe Jarden, and Florian Pop,** Projective group structures as absolute Galois structures with block approximation, 2007

883 **Apostolos Beligiannis and Idun Reiten,** Homological and homotopical aspects of torsion theories, 2007

882 **Lars Inge Hedberg and Yuri Netrusov,** An axiomatic approach to function spaces, spec tral synthesis and Luzin approximation, 2007

881 **Tao Mei,** Operator valued Hardy spaces, 2007

880 **Bruce C. Berndt, Geumlan Choi, Youn-Seo Choi, Heekyoung Hahn, Boon Pin Yeap, Ae Ja Yee, Hamza Yesilyurt, and Jinhee Yi,** Ramanujan's forty identities for Rogers-Ramanujan functions, 2007

879 **O. García-Prada, P. B. Gothen, and V. Muñoz,** Betti numbers of the moduli space of rank 3 parabolic Higgs bundles, 2007

878 **Alessandra Celletti and Luigi Chierchia,** KAM stability and celestial mechanics, 2007

877 **María J. Carro, José A. Raposo, and Javier Soria,** Recent developments in the theory of Lorentz spaces and weighted inequalities, 2007

876 **Gabriel Debs and Jean Saint Raymond,** Borel liftings of Borel sets: Some decidable and undecidable statements, 2007

875 **C. Krattenthaler and T. Rivoal,** Hypergéométrie et fonction zêta de Riemann, 2007

874 **Sonia Natale,** Semisolvability of semisimple Hopf algebras of low dimension, 2007

873 **A. J. Duncan,** Exponential genus problems in one-relator products of groups, 2007

872 **Anthony V. Geramita, Tadahito Harima, Juan C. Migliore, and Yong Su Shin,** The Hilbert function of a level algebra, 2007

871 **Pascal Auscher,** On necessary and sufficient conditions for L^p-estimates of Riesz transforms associated to elliptic operators on $\mathbb{R}^n$ and related estimates, 2007

870 **Takuro Mochizuki,** Asymptotic behaviour of tame harmonic bundles and an application to pure twistor D-modules, Part 2, 2007

869 **Takuro Mochizuki,** Asymptotic behaviour of tame harmonic bundles and an application to pure twistor D-modules, Part 1, 2007

868 **Gelu Popescu,** Entropy and multivariable interpolation, 2006

867 **Vilmos Totik,** Metric properties of harmonic measures, 2006

866 **William Craig,** Semigroups underlying first-order logic, 2006

865 **Nathanial P. Brown,** Invariant means and finite representation theory of $C*$-algebras, 2006

TITLES IN THIS SERIES

864 **John M. Lee,** Fredholm operators and Einstein metrics on conformally compact manifolds, 2006

863 **M. Lübke and A. Teleman,** The Universal Kobayashi-Hitchin correspondence on Hermitian manifolds, 2006

862 **Alberto Canonaco,** The Beilinson complex and canonical rings of irregular surfaces, 2006

861 **Leon A. Takhtajan and Lee-Peng Teo,** Weil-Petersson metric on the universal Teichmüller space, 2006

860 **Thomas M. Fiore,** Pseudo limits, biadjoints and pseudo algebras: Categorical foundations of conformal field theory, 2006

859 **N. Arcozzi, R. Rochberg, and E. Sawyer,** Carleson measures and interpolating sequences for Besov spaces on complex balls, 2006

858 **Enrico Valdinoci, Berardino Sciunzi, and Vasile Ovidiu Savin,** Flat level set regularity of p-Laplace phase transitions, 2006

857 **Donatella Danielli, Nocola Garofalo, and Duy-Minh Nhieu,** Non-doubling Ahlfors measures, perimeter measures, and the characterization of the trace spaces of Sobolev functions in Carnot-Carathéodory spaces, 2006

856 **Vladimir Bolotnikov and Harry Dym,** On boundary interpolation for matrix valued Schur functions, 2006

855 **Yevgenia Kashina, Yorck Sommerhäuser, and Yongchang Zhu,** On higher Frobenius-Schur indicators, 2006

854 **Noam Greenberg,** The role of true finiteness in the admissible recursively enumerable degrees, 2006

853 **Joachim Krieger,** Stability of spherically symmetric wave maps, 2006

852 **Viorel Barbu, Irena Lasiecka, and Roberto Triggiani,** Tangential boundary stabilization of Navier-Stokes equations, 2006

851 **Jie Wu,** On maps from loop suspensions to loop spaces and the shuffle relations on the Cohen groups, 2006

850 **Siegfried Echterhoff, S. Kaliszewski, John Quigg, and Iain Raeburn,** A categorical approach to imprimitivity theorems for C^*-dynamical systems, 2006

849 **Katsuhiko Kuribayashi, Mamoru Mimura, and Tetsu Nishimoto,** Twisted tensor products related to the cohomology of the classifying spaces of loop groups, 2006

848 **Bob Oliver,** Equivalences of classifying spaces completed at the prime two, 2006

847 **Eric T. Sawyer and Richard L. Wheeden,** Hölder continuity of weak solutions to subelliptic equations with rough coefficients, 2006

846 **Victor Beresnevich, Detta Dickinson, and Sanju Velani,** Measure theoretic laws for lim–sup sets, 2006

845 **Ehud Friedgut, Vojtech Rödl, Andrzej Ruciński, and Prasad V. Tetali,** A Sharp threshold for random graphs with a monochromatic triangle in every edge coloring, 2006

844 **Amadeu Delshams, Rafael de la Llave, and Tere M. Seara,** A geometric mechanism for diffusion in Hamiltonian systems overcoming the large gap problem: Heuristics and rigorous verification on a model, 2006

843 **Denis V. Osin,** Relatively hyperbolic groups: Intrinsic geometry, algebraic properties, and algorithmic problems, 2006

842 **David P. Blecher and Vrej Zarikian,** The calculus of one-sided M-ideals and multipliers in operator spaces, 2006

For a complete list of titles in this series, visit the
AMS Bookstore at **www.ams.org/bookstore/**.